Springer Proceedings in Mathematics & Statistics

Volume 279

Springer Proceedings in Mathematics & Statistics

This book series features volumes composed of selected contributions from workshops and conferences in all areas of current research in mathematics and statistics, including operation research and optimization. In addition to an overall evaluation of the interest, scientific quality, and timeliness of each proposal at the hands of the publisher, individual contributions are all refereed to the high quality standards of leading journals in the field. Thus, this series provides the research community with well-edited, authoritative reports on developments in the most exciting areas of mathematical and statistical research today.

More information about this series at http://www.springer.com/series/10533

János D. Pintér · Tamás Terlaky
Editors

Modeling and Optimization: Theory and Applications

MOPTA, Bethlehem, PA, USA, August 2017,
Selected Contributions

Editors
János D. Pintér
Department of Industrial and Systems Engineering
Lehigh University
Bethlehem, PA, USA

Tamás Terlaky
Department of Industrial and Systems Engineering
Lehigh University
Bethlehem, PA, USA

ISSN 2194-1009 ISSN 2194-1017 (electronic)
Springer Proceedings in Mathematics & Statistics
ISBN 978-3-030-12121-1 ISBN 978-3-030-12119-8 (eBook)
https://doi.org/10.1007/978-3-030-12119-8

Library of Congress Control Number: 2018968105

Mathematics Subject Classification (2010): 49-06, 49Mxx, 65Kxx, 90-06, 90Bxx, 90Cxx

This Springer imprint is published by the registered company Springer Nature Switzerland AG
The registered company address is: Gewerbestrasse 11, 6330 Cham, Switzerland

Preface

This volume presents a peer-reviewed selection of works based on lectures given at the Modeling and Optimization: Theory and Applications (MOPTA) 2017 Conference. The conference was held at Lehigh University in Bethlehem, Pennsylvania, United States, between August 16 and 18, 2017.

Following the successful tradition of MOPTA conferences (held every year since 2000), the 2017 meeting brought together researchers and practitioners working on various theoretical and/or practical aspects of optimization model development and solution techniques. Our long-term objective has been to invite presentations that address an interesting range of topics, at the same time providing an ideal setting that enables interaction among the conference participants.

The presentations given at MOPTA 2017 discussed a broad spectrum of topics including models and algorithms to handle challenging combinatorial, linear and nonlinear (both convex and global) optimization problems. A range of optimization challenges arising, e.g., in aircraft design, chemical engineering, energy systems, finance, healthcare, machine learning, multi-objective decision-making, network flows, space engineering, and other prominent research areas were also discussed.

The nine chapters included in this volume offer an interesting cross section of modeling and optimization topics, theory, and intriguing applications, and so they illustrate the diversity of ideas discussed at the conference. Next, we briefly highlight these chapters (following the alphabetical order of their first author): please consult the cited works for in-depth discussions.

Cojocaru, Thommes, and Gillies present an interesting model for residential mail delivery by drones. Defourny and Tu analyze the impact of uncertain gas network disruptions on dual-firing power generation. Fasano and Pintér propose an efficient piecewise linearization approach to handle a difficult class of non-convex optimization problems. Fercoq and Richtárik discuss a smooth minimization method for handling non-smooth functions via parallel coordinate descent methods. Góez and Anjos present second-order conic optimization formulations for service system design problems with congestion. Mut and Terlaky prove that the iteration-complexity upper bound for the Mizuno-Todd-Ye predictor-corrector algorithm is tight. Pirhooshyaran and Snyder introduce a multistage stochastic programming model to optimize the production and distribution network of hip and knee joint replacements. Takáč, Ahipaşaoğlu, Cheung, and Richtárik propose a novel topic discovery algorithm for unlabeled images based on the so-called bag-of-words framework. Wang presents a new exact algorithm to solve non-separable concave quadratic integer programming problems. This brief synopsis of the chapters indicates the truly impressive range of optimization problems that can be handled by innovative optimization models and algorithms.

We are very grateful to the following organizations for partially sponsoring the MOPTA 2017 conference: AIMMS, SAS, Gurobi Optimization, and SIAM. Our special thanks go to the anonymous reviewers of the contributed chapters. We thank Lehigh University for hosting the event, and all other members of the organizing committee—Frank Curtis, Boris Defourny, Ted K. Ralphs, Katya Scheinberg, Lawrence V. Snyder, Martin Takáč, and Luis F. Zuluaga—for their contributions to the success of the conference.

Bethlehem, PA, USA
October 2018

János D. Pintér
Tamás Terlaky

Contents

A Model of Residential Mail Delivery by Drones

Monica G. Cojocaru, Edward W. Thommes and Sierra Gillies

Abstract This paper proposes and analyzes a model of mail delivery in mid-size Canadian urban areas consisting of retrofitting current delivery trucks with a limited number of drones, which can be flown to a number of addresses in a given urban area. Such a truck would then travel between specific locations (called truck-stops), where drone deployment would be executed from. The designated truck-stops are outputs of a proposed algorithm that uses delivery demand data and drone characteristics; they change depending on the needed delivery coverage on a given day. We discuss whether this delivery method could be used in different types of urban areas, where the time to delivery to customers can be shortened, as compared to classic door-to-door delivery.

Keywords Mail delivery · Drones · Logistics of truck with drones · Time to delivery estimates

1 Introduction

The progression of technology has led to the introduction of unmanned aerial vehicles (UAV) also known as drones. Military drone applications are widely known, but numerous civilian applications have also been developed, and they include analyzing atmospheric conditions [5], search and rescue missions [9, 10], post-disaster assessment, environmental management, monitoring infrastructure development [11], image collection [16], traffic surveillance [7], and in proactive weed management [23]. The cost of drones and their power efficiency was also examined

M. G. Cojocaru (✉) · E. W. Thommes · S. Gillies
Department of Mathematics & Statistics, University of Guelph, Guelph, ON N1G 2W1, Canada
e-mail: mcojocar@uoguelph.ca

E. W. Thommes
e-mail: ethommes@uoguelph.ca

J. D. Pintér and T. Terlaky (eds.), *Modeling and Optimization: Theory and Applications*, Springer Proceedings in Mathematics & Statistics 279,
https://doi.org/10.1007/978-3-030-12119-8_1

in [8]. Amazon has introduced a new delivery method involving drones, Amazon Prime Air, guaranteeing the delivery of packages in 30 min or less [26].

Due to the various types of drones today, there have been developments in tools to decide which drone to use in a specific situation from a design state-space viewpoint [2]. The control systems of the drones advanced to become speech-based [22], as well as to incorporate sensors that eliminate the need to remote control them or to provide them with base stations [15]. These control systems are advanced enough to use marker recognition techniques (i.e., estimation of the distance between a marker and the drone by calculating area of recognized marker image), while remaining affordable [17]. Research has also been done to allow drones to have situational awareness during vertical flight [3].

In terms of drones delivering goods and packages, there has been some research done in the areas of path planning and networking. Path planning was studied in [18], using tracking control. Similarly, [14] studied path planning in large urban areas as a continuous space taking into account obstacles. Placement of charging stations was another factor examined in this study. Systems of UAV and UAV bases have been studied more theoretically in [20] to deal with shipping of goods on a regional scale, and beyond urban deliveries. On the other hand, in [12], a mobile ad-hoc network is examined with drones that can communicate. The operational requirements of the drones lead to advanced networking requirements. Specific package delivery landing platforms were designed in [21]. Considerations include where to efficiently position a landing platform, as well as how to handle situations where packages are redirected to more accessible locations. Along with redirecting the packages, a communication system between the drones and the platform was designed which was also used to ensure that the delivery was executed. In [29] the authors examined scheduling methods for UAV interregional delivery systems, which were optimized to reduce time until delivery. Systems are commanded/controlled from regional UAV centers. They included weight-based scheduling and an optimal allocation scheme by using a dynamic programming approach. In [30], the authors provided a system for mail delivery with drones where the drones can communicate with external devices. These drones also have anti-theft protection in the form of authorized access to goods.

In this paper, we propose a model in which a delivery truck stops at designated addresses and, from there, uses drones to deliver parcels in a specified surrounding subarea, thus potentially reducing delivery time and costs involved in current door-to-door (DTD) methods. Our model proposes that packages be delivered door-to-door by drones, without eliminating the jobs of truck drivers. The proposed model is more efficient than the current door-to-door method in terms of time to delivery, and we show under which conditions this statement holds. We show that the radius of flight of the drones, the speed of the drones, and the number of drones to be hosted in a truck should not be chosen independently, since they all affect the proposed delivery method. The closest paper to date to our own is the [19], where the authors propose a truck delivery with a drone "sidekick", i.e., a truck and driver current system with one drone that can execute a number of deliveries concurrent with the driver. The work we propose differs from [19] in several ways, in the sense that the truck can carry multiple drones and can execute simultaneous deliveries from one location, and in that, justifiably, the driver's delivery schedule can be optimized between truck-stops using an exact solution to the traveling salesman problem, given that the number of truck-stops is small. We also offer a computational method for outlining the time to delivery comparisons between the DTD model and the proposed truck-with-drones (TwD) model.

The outline of the paper is as follows: Sect. 2 presents our motivation and our model description. We outline an easily implementable algorithm to select optimal truck-stops from which drones can deliver packages within a given day. Section 2.3 presents a time to delivery estimate between our model and the traditional door-to-door model, while Sect. 3 shows the sensitivity of TwD to parameters depending on type of drones, truck carrying capacity for drone units, demand for delivery, and size of urban area to deliver to over a month (30 days) delivery period. We present our conclusions in Sect. 4.

2 Model Description

2.1 Motivation

Current by mail delivery services (such as FedEx, Canada Post, UPS, etc.) involve consumers who either have to travel to a post office to pick up their package, or have their package delivered to the door. Our view in this paper is that a delivery company could use advanced technology in mail/package delivery as an add-on service. We thus propose and analyze a model where regular delivery trucks can be outfitted with a number of drones that the driver can deploy to nearby addresses in urban areas (which we discuss in detail in the next sections). We show that our proposed logistic framework might be suited for certain urban areas and certain densities of delivery addresses. We show that time to delivery can decrease, while current resources (such as drivers and trucks) can be used, assuming customers would agree to sign on to such a service.

A customer sign-on/user fee could offset the price of acquiring and operating the UAV, however a cost–benefit analysis is beyond the scope of our research here. Instead, we show how the type of urban areas and demand of delivery are linked with shortening time to delivery between our proposed truck-with-drones model and a regular door-to-door service. The main urban area in our model is inspired by a mid-size urban area of Southwestern Ontario, specifically the city of Guelph, which has 125,000 inhabitants and encloses roughly a 10×10 km area. We considered Guelph a representative mid-size suburban area in Ontario, and we compounded and used data for times and distances on the road within city limits using Google maps. The distance matrices were used to inform our simulated distances below.

2.2 Delivery Scheme and the Truck-Stop Algorithm

Instead of delivering DTD, a truck could be adjusted to carry a number of drones. Assuming that we have N addresses to be delivered to in a given time interval, we would like to solve the following problem: find a delivery address containing the most other delivery addresses within a predefined radius around it. The predefined radius depends on drone characteristics, specifically velocity and battery charge. We call this address a "truck-stop". Once the problem is solved for N addresses, we pose the same problem for the remaining addresses obtained by removing the previously discovered truck-stop and its associated addresses from N. Continuing iteratively, we are able to re-list all N delivery addresses in the form of individual truck-stops and their associated addresses. Thus, we propose an algorithmic approach to the problem of re-listing a given number of delivery addresses N in this manner. In mathematical formalism:
Step 0: Initiate N addresses in an area by their (x_i, y_i), $i \in \{1, \ldots, N\}$ coordinates; the radius is considered fixed, depending on the drone model; we generically denote it R.
Step 1:

1. Find the first address $i_1 \in \{1, \ldots, N\}$ with the most addresses within R around its coordinates (x_{i_1}, y_{i_1}). This is the first truck-stop and we denote by m_1 the number of addresses around i_1.
2. Record the list of all m_1 addresses;

Step k:

1. Repeat *Step* 1 on the remaining $\{1, \ldots, N\}\backslash\left(\bigcup_{l=1}^{k-1} m_l\right)$;
2. Record the subset m_k containing the k-th truck-stop and its addresses within the given radius.

End Step:

1. Repeat *Step k* until all addresses are exhausted. List all subsets m_k found in steps above.
2. Optimize the route among all truck-stops by solving the TSP problem among truck-stops listed.
3. List all addresses not recorded in any of the previous Steps; these are called *leftover addresses*; they will be covered in normal door-to-door driver delivery, thus they will not contribute to reducing time to delivery in the area.

An address i as described above is a *truck-stop*. The algorithm gives a list of *truck-stops* for the truck driver to stop at, together with the addresses in each truck-stop. The driver then is required to follow a route through the truck-stops in an optimal way. Started in 2013, and completed in 2017, UPS for instance uses an algorithm called ORION (On-Road Integrated Optimization and Navigation) [24] to solve this problem. ORION analyzes a collection of data points including the day's package

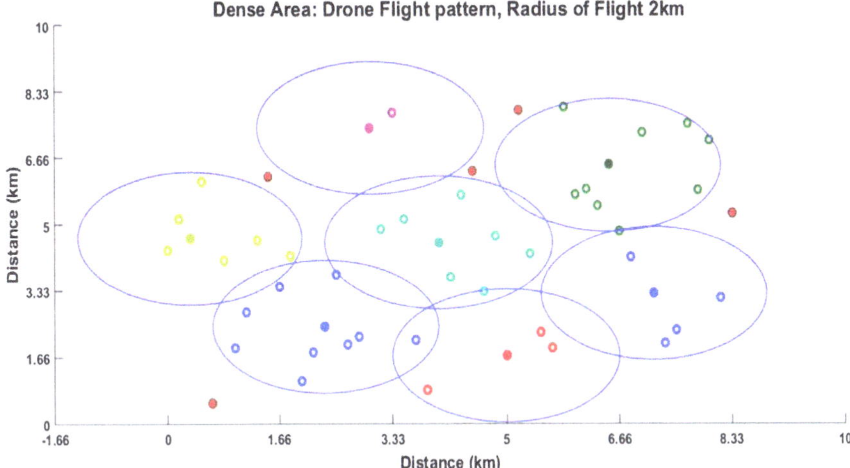

Fig. 1 Truck-stops algorithm for 51 addresses. The circles correspond to deliveries within the same stop—blue square truck-stops. The few red filled-in circles are additional addresses outside the radius of all others, hence they are done in a door-to-door manner. Here, a drone radius of flight was considered to be 2 km

deliveries, pickup times, and past route performance to create the most efficient daily route for drivers. In contrast, we model our route among the truck-stops as a traveling salesman problem (TSP).

We implemented this algorithm in Matlab R2015a on a Windows 10 Microsoft Surface Book, Windows 10 Pro 64 bit with Intel(R) Core(TM) i7-6600U CPU. On average, the run time for our code would be of an order of 10 s for up to $N = 100$ addresses in the area. A truck driver drives from truck-stop to truck-stop, as shown for instance in Fig. 1, and at each stop, they use drones to deliver in that area. The algorithm records the truck-stops and the optimal route (see Fig. 2).

2.3 Time to Delivery in the Truck-with-Drones Model

In this section, we expand the analysis of our delivery model by adding a back of the envelope estimate of the upper bound of time to delivery to customers from a given truck-stop. According to Sect. 2, a given urban area is divided into a number of subareas based on the total daily number of addresses, N, to be delivered to in a day. Then, the algorithm in Sect. 2.2 lists the delivery addresses for each truck-stop, where a truck-stop delimits a delivery subarea. Once a driver arrives at a truck-stop, they load packages into drones which then take flight (simultaneously) to deliver packages. If there are more addresses to deliver to from a given truck-stop than there are drones in the truck, the drones must make multiple trips, which we call deployments. So once the drones return, they are reloaded (with both parcels and

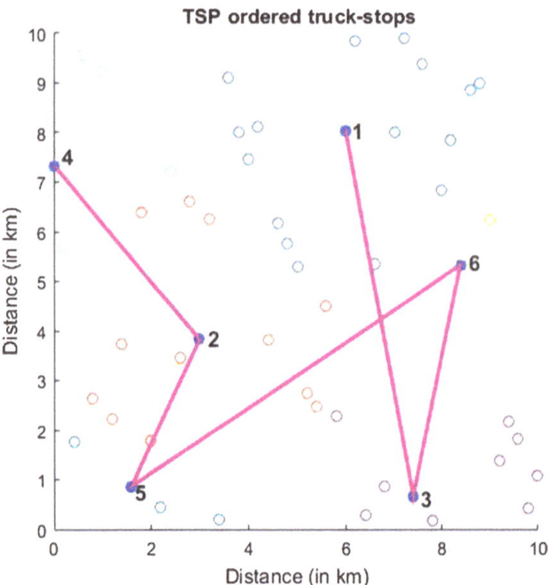

Fig. 2 Truck-stops algorithm for 51 addresses with a drone radius of flight of 2 km. The optimal route between the 6 truck-stops is shown

fresh battery packs if needed) and sent out again. The drones will continue to make trips until deliveries to all addresses around the truck-stop are exhausted.

To set our ideas further, let us denote by $\{1, \ldots, S\}$ the number of delivery subareas in a day, and by N_d the number of drones in the truck. Let $k \in \{1, \ldots, S\}$ be a generic subarea around the i_k truck-stop and let $N_k < N$ be the number of delivery addresses within the subarea k. Since i_k is the truck-stop, then all other addresses $j \in N_k \backslash \{i_k\}$ fall within one drone deployment radius (R_{max}) from i_k, i.e., we further assume for the time-being that $N_k \leq N_d$.

The time to make deliveries from i_k to all other addresses j, given simultaneous drone deliveries at all j, is calculated by determining the maximum round-trip distance from the truck-stop to any of the j addresses, denoted by $d_{i_k}^j$, then dividing it by the speed of the drone v_d (we assume flight in a straight line). We thus have:

$$t_k = \leq \frac{d_k^{max}}{v_d} := \frac{\max_j d_{i_k}^j}{v_d}, \text{ for each } k \in \{1, \ldots, S\}. \tag{1}$$

Note that the right-hand side of (1) is an overestimate of the time to delivery in a subarea, since we consider all drone trips to take as long as traveling to furthest address from i_k, within subarea k.

We can generalize our estimate (1) by dropping the assumption that $N_k \leq N_d$. If $N_k > N_d$, then multiple deployments may occur from the same truck-stop i_k in order to cover all addresses in the k-th subarea. The number of deployments necessary is given by $\left[\frac{N_k}{N_d}\right] + 1$, where by $[x]$ here we denote the floor function of x. Then (1) becomes:

$$t_k = \left(\left\lceil \frac{N_k}{N_d} \right\rceil + 1\right)\frac{d_k^{max}}{v_d} + \Delta t \left\lceil \frac{N_k}{N_d} \right\rceil, \text{ for } k \in \{1, \dots, S\}, \tag{2}$$

where Δt stands for an estimate of reloading time for all drones between two consecutive deployments, while $\left\lceil \frac{N_k}{N_d} \right\rceil$ is the number of reloads.

Given that we are searching for conditions under which a TwD model is more time efficient than a DTD model, we see here that the reload time for drones may play an important role, in addition to optimizing driver's routes and truck-stop numbers. From above, we see that in the subarea k, if we desire $t_k < t_{DTD}$ (where by t_{DTD} we denote the time to delivery in k via the traditional DTD model optimized with a TSP route), we have to have that:

$$\Delta t < \frac{t_{DTD} - \left(\left\lceil \frac{N_k}{N_d} \right\rceil + 1\right)\frac{d_k^{max}}{v_d}}{\left\lceil \frac{N_k}{N_d} \right\rceil}. \tag{3}$$

This seems to indicate that TwD is more efficient than the traditional DTD if Δt satisfies the last inequality; otherwise, it may not be.

The delivery times from each subarea are summed up to give a total delivery time for a day:

$$T_{day} = \sum_{k=1}^{S} t_k + T_{stops}, \text{ where } T_{stops} = \text{TSP travel time between truck-stops.} \tag{4}$$

There are two important remarks to highlight at this point.

Remark 1 (a) To compute T_{stops}, we solve a traveling salesman problem (TSP) with all truck-stops being the nodes the driver needs to visit. For an urban area such as Guelph, Ontario, given our algorithm above, we find that the number of truck-stops is ≈6–8, which makes the solvability of the TSP possible.
(b) To minimize Δt, it may be possible for instance to implement a delivery scheme where the number of drones in the truck is doubled, and they are staggered in flight. This would mean while one group is making deliveries, the other is charging/being prepared. If more drones take too much space, one could have available extra battery packs readily charged.

3 Sensitivity Analysis of the Delivery Model

In this section, we design and simulate several experiments to outline how the time to delivery estimates proposed in our model so far are affected by varying parameters. We are also able to derive some estimates that shed light on the size of Δt.

3.1 Time to Delivery from a Single Truck-Stop

First, we develop a side-by-side comparison between the traditional door-to-door (DTD) delivery and our TwD model, from the perspective of delivery time to customers *from a single truck-stop*. We analyze scenarios in which there are a numbers of delivery addresses $N \in \{4, 6, 8, 9\}$ all located around a truck-stop. The drone numbers available in a truck, $N_d \in \{2, 4, 6, 8\}$. The drone speed was set to $v_d = 40$ km/h. The truck speed was set at an average of 45 km/h on city roads. We averaged 20 experiments for each pair of (N, N_d) values and we plotted our findings in Fig. 3, where in each experiment, we randomly created the number and locations of the N addresses.

We created and used a distance data file with addresses from Guelph, ON, and recorded their distances using Google maps. We used this distance matrix to randomly assign distances between the N addresses in each experiment.

Using the notations of last subsection, we computed and plotted t_k as in (2) where $k = 1$ corresponding to a single truck-stop. We assumed a $\Delta t = 0$ even for cases with multiple deployments and we plotted the values in Fig. 3 in blue and green columns. For the t_{DTD} estimate, we found a TSP route among all N addresses considered in each case. We plotted our t_{DTD} estimates in Fig. 3 in red columns. Both time estimates are measured in fractions of 1 hour physical time.

We see that the TwD model could shorten the time to delivery to customer when the number of drones in the truck N_d is enough to make a single deployment possible from the truck-stop. These are clearly shown in the green colored columns in 3 of the 4 panels of Fig. 3.

At the other spectrum, we see that the DTD model is more efficient when N_d is small: specifically, even with $\Delta t = 0$, we see that $N_d = 2$ is too small a number of drones to compete with DTD. When $N_d = 4$ and $N \geq 6$ (lower panels of Fig. 3), the TwD may not be competitive with DTD, since either.

As we increase N the likelihood of multiple deployments out of the truck-stop increases, the reload time between deployments, Δt, becomes more important. For $N_d = 6$ and $N = 8$, estimate (3) has to be considered for a proper conclusion to our comparison. Here, we have 1 reload and 2 deployments. According to our values, we need to have $\Delta t < t_{DTD} - t_{drones^1} \approx 0.05 \approx 3$ min, which[1] may be impossible. In the lower right panel, when $N_d = 6$ and $N = 9$, we have more room for $\Delta t < t_{DTD} - t_{drones} \approx 0.1 \approx 6$ min.

Therefore, the number N_d of drones in a truck should not only be chosen with respect to a drone's velocity and radius of flight but also with respect to the density of addresses in need of delivery in given subareas, which in turn determines the number of consecutive deployments from a truck-stop.

It seems that the truck-with-drones model is more time efficient than the door-to-door model in some cases, such as when one deployment out of the truck-stop covers all N, or when N_d is larger. However, it may be unrealistic to expect to carry a large

[1] Here, we take the height difference between red column in lower left panel of Fig. 3 and the height of the $N_d = 6$ blue column in the same panel.

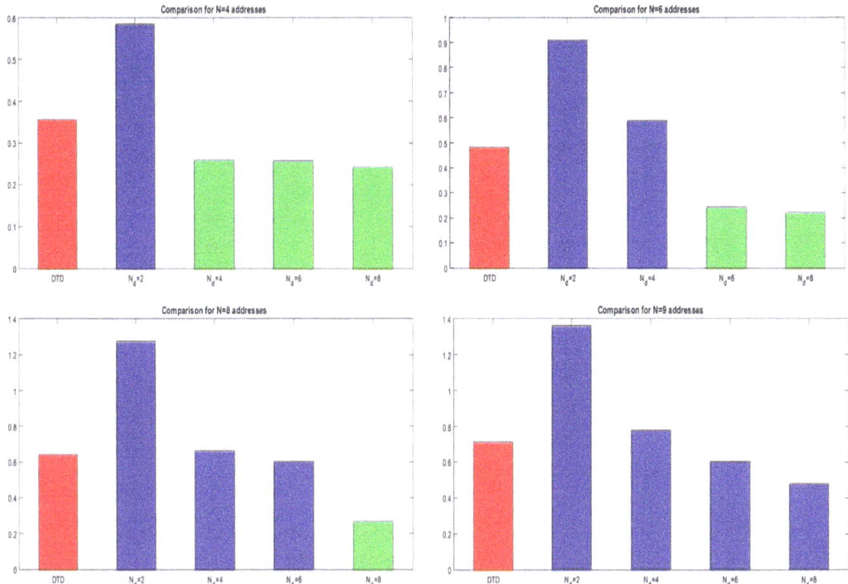

Fig. 3 Delivery times for DTD and truck with drones in U_2, with $N_d \in \{2, 4, 6, 8\}$ and a delivery density of $N = 4$ (upper left), $N = 6$ (upper right), $N = 8$ (lower left) and respectively $N = 9$ (lower right) addresses from one truck-stop. Blue bar times are obtained from multiple deployments; green bars are single deployment. The vertical axis measures fraction of 1 h physical time

number of drones such as $N_d = 8$, as trucks have to also hold the packaged needed to be delivered. Having fewer drones ($N_d = 4$ or 6) may reduce delivery time while remaining more implementable, as less drones take up less space in truck, however in this case, the size of Δt will play a role in determining the feasibility of drone delivery versus DTD.

3.2 TwD Time to Delivery over Large Areas with Varying Demand Levels

Expanding from our previous results, we design and simulate here two more experiments in which we vary the demand level of deliveries (N), the size of the urban area the demand takes place over, as well as drone velocities correlated with maximal radius of flight. Parameter ranges for our experiments can be found in Table 1. We are interested in observing how the flight time portion of the time to delivery (i.e., $\sum_{k=1}^{S} t_k$, as defined in (4)) is affected by the change in drone characteristics, geographic area, and demand levels.

The first experiment varies the drone speed v_d and the number of drones in the truck, $N_d \in \{4, 6, 8\}$, while it selects a random number of delivery addresses $N \in$

Table 1 Parameters and variables examined in the sensitivity analysis of an area with multiple truck-stops

Parameter	Variations	Corresponding parameters
Number of deliveries	Random pick in [10, 40]	
	Random pick in [41, 80]	N
	Random pick in [81, 120]	
Drone speed	v_d <40 km/h	$R_{max} = 7.5$ km
	$v_d \in [40, 60]$ km/h	$R_{max} = 5$ km
	$v_d > 60$ km/h	$R_{max} = 2.5$ km
Number of drones	4	
	6	N_d
	8	
Size of area	10×10 km^2	
	20×20 km^2	

[10, 40] per area, for 30 days. Velocity ranges for the drones are in Table 1. The results are shown in Fig. 4 where each day plots the average values over 20 simulation runs of flight time in a day. In the majority of instances simulated, we observed single deployments per truck-stop, thus we used (1). The number of truck-stops, as depending on v_d and R_{max}, varied in our simulations between 2 and 6, with an average of 4 addresses per truck-stop. The time is in fraction of 1 h.

For an estimate of t_{DTD} in these simulations, we extract from Sect. 3 an average estimate of a TSP route with 4 nodes to be $0.365 \approx 22$ min. For 2 truck-stops scenarios, this means a value of 0.73 and for a 6 truck-stops scenario, this means a value of 2.19. Thus $t_{DTD} \in [0.73, 2.19] \approx [44$ min, 132 min$]$.

In our comments below, we disregard the time it would take the driver to do DTD among the truck-stops, as that time would need to be spent equally in TwD and in DTD models.

We note that, while packing more drones in a truck leads to shorter delivery times, the difference in time savings between the medium and high velocity drones is again not as large as the one between medium and low velocity ones. Overall, we see that the maximal flight time value in Fig. 4 is approximately 0.75, which means that on average, over the 30 days, the TwD model has shorter delivery times compared to $t_{DTD} \in [0.73, 2.19] \approx [44$ min, 132 min$]$.

We note also that having 8 drones does not change delivery times significantly from having 6 drones. This is interesting also from the perspective of the results in Sect. 3 above, where time efficiency seemed to require $N_d \geq 6$. In Fig. 4, there is little difference between medium and high velocity drones; since velocity correlates to the flight radius R_{max}, then the conclusion of the sensitivity analyses so far is that a truck with $N_d \geq 6$ and medium velocity will achieve a shorter time to delivery than traditional TDT model. In our last experiment, we further refine our simulated scenarios to try and identify under what circumstances, a truck carrying a higher number of drones would lead to further time savings. We run an experiment with $N_d \in \{6, 8\}$ drones with all three types of velocity v_d, over a larger area with a variable

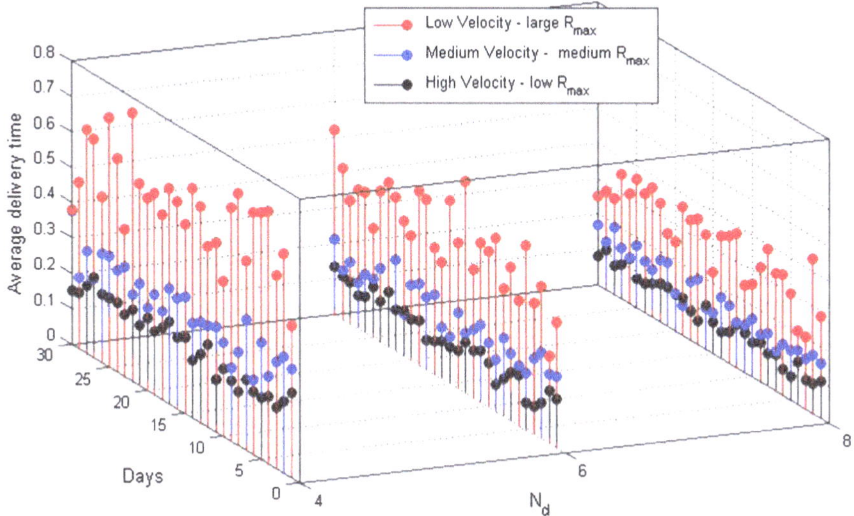

Fig. 4 Average delivery times over a 30 days simulation for different v_d and R_{max}, as well as different N_d values. Vertical axis presents time estimates measured in fractions of 1 h physical time

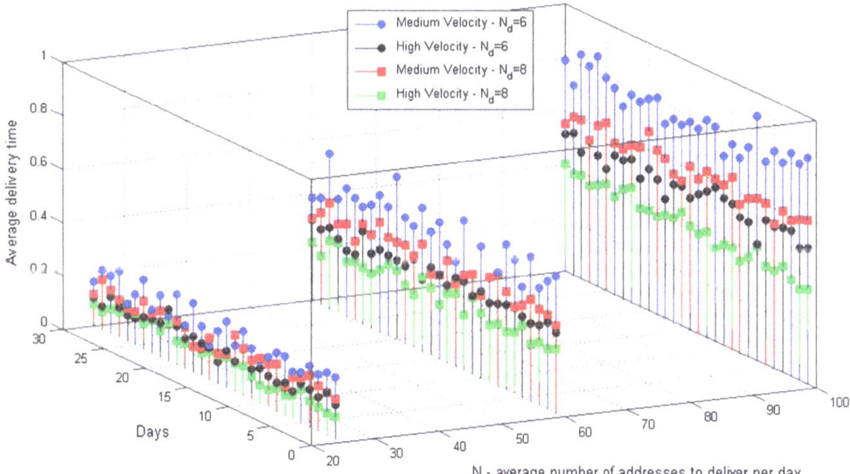

Fig. 5 Average delivery times over 30 days for different drone speeds for a larger area with $N_d \in \{6, 8\}$. The density of deliveries N is Low: $N \approx 25$, $Medium$: $N \approx 60$ and respectively $High$: $N \approx 100$. Vertical axis presents time estimates measured in fractions of 1 h physical time

delivery demand, N. We simulate a $20 \times 20\,\text{km}^2$ area (shown in Fig. 5), where N was randomly picked between [10, 40], [41, 80], respectively [81, 120] deliveries per day (each day's results were averaged over 20 simulation runs). The delivery densities are referred to in the graph as Low: $N \approx 25$, $Medium$: $N \approx 60$ and respectively $High$: $N \approx 100$.

We see that an increase in deliveries per day leads to an increase in delivery time. The second conclusion is that "more is better" in certain cases: using more drones in a truck $N_d = 8$ and using the fastest ones (green dots in Fig. 5) is most efficient during a high demand of delivery. At low demand, there is little difference made by either the number of drones N_d or their velocity.

This would suggest that there is a clear dependency between potential time savings on delivery and the demand for deliveries. Perhaps retrofitting several truck sizes with various numbers of drones would be more appropriate. Larger trucks can be used in peak demand periods—such as Christmas time, Easter time, etc.—while smaller trucks can be used in usual/low demand periods.

4 Further Discussions and Conclusions

Whether implementing drones for mail delivery would be more time efficient than current door-to-door depends on the demand for delivery (which in part may depend on the demographics of the area) and on the number and type of drones considered. For the truck delivery model proposed, this would mean paying attention to the relation between the carrying capacity of the truck, for both packages and drones, and demand for deliveries. Upgrading trucks to fit the same number of packages but with added space for carrying the drones would come at a cost, and to this cost a company needs to add the costs of drone operators (retraining drivers) that need to be employed. The cost of drones themselves depends on the type considered.[2] On the other hand, this model means that trucks may drive less on city roads (decreasing pollution levels, avoiding road congestion etc.), while possibly achieving better delivery times to customers.

So far, our analysis led us to conclude that, given roughly one deployment per truck-stop in a given subarea, then time to delivery can be improved if a truck is retrofitted with anywhere between $N_d \in [4, 8]$ drones for daily delivery numbers of $N \in [10, 100]$. Moreover, the algorithm we propose to compute truck-stops (in Sect. 2.2) can be easily modified to output a list of truck-stops, each with a maximum number of delivery addresses within R_{max} equal to N_d, thus assuring that the driver executes exactly one deployment of drones at each stop.

In general, any urban area has locations where multiple packages are needed to be delivered at the same time. This includes places such as apartment buildings, office suites, etc. The drones in our model would have to take multiple trips from a truck to deliver these individually, which is inefficient, as it increases the number of deployments. Thus, in areas where delivery density is consistently high, installation of drone depots could be another delivery scheme. The depots would include multiple drones and charging stations, and as packages arrive they would be immediately

[2]For instance, we found an estimated cost for Amazon prime of \$0.88 per delivery or of \$0.24 for a delivery of a 2 kg package over 10 km, but the latter drones [26] also had docking stations which would not be needed here.

flown out. The drone depot can be modeled as the source location as in [29], where a weight-based scheduling scheme takes into account the delivery distance and the priority of the delivery package. This method minimized the probability of delay of delivery while maximizing the ability of UAVs to successfully handle incoming requests.

New delivery schemes involving drones can be considered depending also on the demographics of an urban subarea. If there is a large demand of packages or the area is denser, such as in a student living area, a drone depot may be more useful. However, if areas are less dense and include more single dwelling residential homes and schools, such as the North end of Guelph, the truck delivery model may be less costly to implement than a depot station, while still being more efficient than the current door-to-door model. Last but not least, the radius of flight may be an interesting factor. It may be more useful in the truck model to have a larger radius but fewer drones, whereas in the depot model, a smaller radius with lots of drones could be used.

Drone failure rates were unavailable for this study but should be taken into account when considering implementation of a drone delivery model, though it is likely that losses in content and/or equipment can be mitigated via insurance policies. Along with potential benefits, there are concerns regarding drone use. Examples are loss of control during flight, risk of drone battery depletion [4], and systems health and capability degradation during long duration flights [1]. Finally, mitigating privacy issues would be another drawback to implementing such a service, as many drone models are able to carry cameras that record their flight and their surroundings. Ethical concerns, including privacy, arise with the commercial use of drones [27]. Autonomous drones capable of causing fatalities are also a concern [27]. Currently, the FAA is working to implement regulations for flying autonomous drones. All these considerations should be taken into account, alongside financial gains, by any company wishing to implement a drone component to their delivery operations.

4.1 Future Work

An expansion of this work would be a side-by-side comparison of out time estimates and of the truck with sidekick model of [19]. Considerations and models of implementation costs and benefits would improve our analysis here. This analysis would touch two facets regarding a possible real world implementation: the costs would directly impact the price for this service at consumer level, while benefits could contribute to the greater good, such as reduced traffic congestion, cleaner air, etc.

We started this work a couple of years ago, and it seems we are on the right path. To see a partial (and unrelated to our work) real-life implementation of some of the ideas, see [25].

4.2 Data

All data on cities and their demographics stated above was collected via Google Maps. The door-to-door delivery time in the South end area of Guelph was collected using Google Map's directions feature. Estimations of densities of delivery were from online interviews with truck drivers. The truck-stop algorithm can be implemented with address data from an existing data base of postal codes and the respective distances between them.

Acknowledgements This work was supported by the National Science and Engineering Research Council of Canada (NSERC) under Discovery Grant [number 400684] of first author, under Discovery Grant [number 400551] of the 2nd author and an NSERC USRA award for undergraduate students (3rd author).

References

1. Agha-Mohammadi, A.A., Ure, N.K., How, J.P., Vian, J.: Health aware stochastic planning for persistent package delivery missions using quadrotors. In: IEEE International Conference on Intelligent Robots and Systems (IROS 2014), pp. 3389–3396 (2014). https://doi.org/10.1109/IROS.2014.6943034
2. Ali, A., Ballou, N., McDougall, B., Valle Ramos, J.L.: Decision-support tool for designing small package delivery aerial vehicles (DST-SPDAV). In: Systems and Information Engineering Design Symposium (SIEDS), pp. 45–50 (2015)
3. Astrov, I., Pedai, A.: Situational awareness based flight control of a drone. In: IEEE International Systems Conference (SysCon) 2011, pp. 574–578 (2011). https://doi.org/10.1109/SYSCON.2011.5929033
4. Berenz, V., Suzuki, K.: Risk and gain battery management for self-docking mobile robots. In: IEEE International Conference on Robotics and Biomimetics (ROBIO), pp. 1766–1771 (2011). https://doi.org/10.1109/ROBIO.2011.6181545
5. Berman, E.S.F., Fladeland, M., Liem, J., Kolyer, R., Gupta, M.: Greenhouse gas analyzer for measurements of carbon dioxide, methane, and water vapor aboard an unmanned aerial vehicle. Sens. Actuators B Chem. **169**(2012), 128–135 (2012)
6. Canada Post's Five Point Action Plan. http://www.canadapost.ca/cpo/mc/assets/pdf/aboutus/5_en.pdf (2015)
7. Chen, Y.M., Dong, L., Oh, J.-S.: Real-time video relay for UAV traffic surveillance systems through available communication networks. IEEE Wirel. Commun. Netw. Conf. WCNC 2007, 2608–2612 (2007)
8. D'Andrea, R.: Guest editorial: can drones deliver?, IEEE Trans. Autom. Sci. Eng. **11**(3), 647–648 (2014)
9. Doherty, P., Rudol, P.: A UAV search and rescue scenario with human body detection and geolocalization. In: AI 2007: Advances in Artificial Intelligence. Lecture Notes in Computer Science, vol. 4830, pp. 1–13 (2007). https://doi.org/10.1007/978-3-540-76928-6_1
10. Erdos, D., Erdos, A. and Watkins, S.E.: An experimental UAV system for search and rescue challenge. IEEE Aerosp. Electron. Syst. Mag. **28**(5), 32–37 (2013)
11. Ezequiel, C.A.F., Cua, M., Libatique, N.C., Tangonan, G.L., Alampay, R., Labuguen, R.T., Favila, C.M., Honrado, J.L.E., Canos, V., Devaney, C., Loreto, A.B., Bacusmo, J., Palma, B.: UAV aerial imaging applications for post-disaster assessment, environmental management and infrastructure development. In: International Conference on Unmanned Aircraft Systems (ICUAS), 2014, pp. 274–283 (2014). https://doi.org/10.1109/ICUAS.2014.6842266

12. Frew, E.W., Brown, T.X.: Networking issues for small unmanned aircraft systems. J. Intell. Robot. Syst. **54**(1–3), 21–37
13. Google maps API for route optimization. http://www.optimap.net/
14. Hong, I., Kuby, M., Murray, A.: Deviation flow refueling location model for continuous space: commercial drone delivery system for urban area. In: Proceedings of the 13th International Conference on Geocomputation Geospatial Information Sciences (2015)
15. Jimenez Lugo, J., Zell, A.: Framework for autonomous onboard navigation with the AR.Drone. In: International Conference on Unmanned Aircraft Systems (ICUAS), pp. 575–583 (2013). https://doi.org/10.1109/ICUAS.2013.6564735
16. Johnson, L.F., Herwitz, S., Dunagan, S., Lobitz, B., Sullivan, D., Slye, R.: Collection of ultra high spatial and spectral resolution image data over California vineyards with a small UAV. In: Proceedings of the 30th International Symposium on Remote Sensing of Environment, vol. 20, pp. 845–849 (2003)
17. Kim, J., Lee, Y.S., Han, S.S., Kim, S.H., Lee, G.H., Ji, H.J., Choi, H.J., Choi, K.N.: Autonomous flight system using marker recognition on drone. In: Frontiers of Computer Vision (FCV), 21st Korea-Japan Joint Workshop, pp. 1–4 (2015). https://doi.org/10.1109/FCV.2015.7103712
18. Kim, H.J., Shim, D.H., Sastry, S.: Nonlinear model predictive tracking control for rotorcraft-based unmanned aerial vehicles. In: American Control Conference Proceedings 2002, vol. 5, pp. 3576–3581 (2002)
19. Murray, C.C., Chu, A.G.: The flying sidekick traveling salesman problem: optimization of drone-assisted parcel delivery. Transp. Res. Part C Emerg. Technol. **54**(86–109) (2015)
20. Pfaender, H., DeLaurentis, D., Mavris, D.: An object-oriented approach for conceptual design exploration of UAV-based system-of-systems. In: Proceedings of 2nd AIAA Unmanned Unlimited Conference, Ser. AIAA, vol. 6521 (2003)
21. Sisko, M.: Robotic aerial vehicle delivery system and method, Windermere, FL, US, no. 20150158599. www.freepatentsonline.com/y2015/0158599.html (2015)
22. Supimros, S., Wongthanavasu, S.: Speech recognition—based control system for drone. In: Third ICT International Student Project Conference (ICT-ISPC), pp. 107–110 (2014). https://doi.org/10.1109/ICT-ISPC.2014.6923229
23. Torres-Sánchez, J., López-Granados, F., De Castro, A.I., Peña-Barragán, J.M.: Configuration and specifications of an unmanned aerial vehicle (UAV) for early site specific weed management. PLoS ONE **8**(3), e58210 (2013). https://doi.org/10.1371/journal.pone.0058210
24. UPS Press Release, 2016: UPS Wins 2016 INFORMS Franz Edelman Award for Changing the Future of Package Delivery (n.d.). https://pressroom.ups.com/pressroom/ContentDetailsViewer.page?ConceptType=PressReleases
25. Stewart, J.: A drone-slinging UPS van delivers the future. In: Wired. https://www.wired.com/2017/02/drone-slinging-ups-van-delivers-future/ (2017)
26. Wang, D.: The economics of drone delivery. Online at Flexport.com. https://www.flexport.com/blog/drone-delivery-economics/ (2015)
27. Wilson, R.L.: Ethical issues with use of drone aircraft. In: IEEE International Symposium on Ethics in Science, Technology and Engineering, pp. 1–4 (2014). https://doi.org/10.1109/ETHICS.2014.6893424.
28. Xu, C., Zhang, X.: A routing algorithm for schismatic communication network based on UAV. In: IEEE 4th International Conference on Electronics Information and Emergency Communication (ICEIEC), pp. 49–52 (2013). https://doi.org/10.1109/ICEIEC.2013.6835451
29. Zhang, H., Wei, S., Yu, W., Blasch, E., Chen, G., Shen, D., Pham, K.: Scheduling methods for unmanned aerial vehicle based delivery systems. In: IEEE/AIAA 33rd Digital Avionics Systems Conference (DASC), pp. 6C1–1 (2014)
30. Zhou, T.T.G., Zhou, D.T.X., Zhou, A.H.B.: Unmanned drone, robot system for delivering mail, goods, humanoid security, crisis negotiation, mobile payments, smart humanoid mailbox and wearable personal exoskeleton heavy load flying machine. Google patents, US Patent App. 14/285,659 (2014)

On the Value of Dual-Firing Power Generation Under Uncertain Gas Network Access

Boris Defourny and Shu Tu

Abstract This work is concerned with the impact of gas network disruptions on dual-firing power generation. The question is addressed through the following optimization problem. Markets drive the price of gas, oil, and electricity. The log-prices evolve as correlated mean-reverting processes in discrete time. A generating unit has dual-firing capabilities, here in the sense that it can convert either gas or oil to electricity. Oil and gas are subject to different constraints and uncertainties. Gas is obtained in real time through the gas network. Due to gas supply disruptions, gas access is not guaranteed. Oil is stored locally and available in real time. The oil storage capacity is limited onsite. Oil can be reordered to replenish the oil tank, with a lead time between the order time and the delivery time. Oil is paid for at the order time price. In this paper, we formulate the stochastic optimization problem for a risk-neutral operator, and study the sensitivity of the value of the dual-firing generating unit to the gas network availability parameters.

Keywords Dual-firing · Power generation · Energy asset management · Natural gas-electric coordination · Resilience · Markov decision processes · Optimal control · Dynamic programming · Stochastic optimization

MSC (2010): 90B05 · 90B25 · 90C15 · 90C39 · 90C40

1 Introduction

This paper is concerned with the impact of gas network disruptions on dual-firing power generation. The question is addressed through the following optimization problem. There are three markets to describe the price of gas, oil, and electricity. The

B. Defourny (✉) · S. Tu
Department of Industrial and Systems Engineering, Lehigh University, 200 W Packer Ave, Bethlehem, PA 18015, USA
e-mail: defourny@lehigh.edu

S. Tu
e-mail: sht213@lehigh.edu

J. D. Pintér and T. Terlaky (eds.), *Modeling and Optimization: Theory and Applications*, Springer Proceedings in Mathematics & Statistics 279,
https://doi.org/10.1007/978-3-030-12119-8_2

log-prices evolve as correlated mean-reverting processes in discrete time. A dual-firing power generation unit can convert fuel to electricity. Usually, the prices and heat rates are such that it is more profitable to use gas rather than oil. However, the fuels are subject to different constraints and sources of uncertainty. Gas is physically obtained in real time through a gas pipeline. Due to gas supply disruptions, physical gas access is not guaranteed. The availability of the gas network is described by a two-state Markov chain. The parameters of the transition matrix determine the frequency and mean duration of the gas network disruptions. Oil is stored locally and available in real time. However, the oil storage capacity is limited onsite. Oil can be reordered to replenish the oil tank, but there is a lead time between the order time and the delivery time. Oil is paid for at the order time price.

In this paper, the questions that are raised concern the determination of an optimal fuel utilization and oil replenishment policy, the value of the dual-firing generator asset, and the sensitivity of the value of the generator asset to the parameters of the random process describing the availability of the gas network. Gains in value over a single-fuel generator may be used to justify investments in dual-firing capabilities and oil storage capacity. Intuitively, we expect that the value added by dual-firing increases as the gas network becomes less reliable. To calculate the gain in value however, one must first determine an optimal operations policy. In this paper, we formulate the problem as a Markov decision process in a continuous state space with unbounded rewards. We examine its relationship with an equivalent multistage stochastic programming formulation. As we cannot solve these problems exactly, we resort to lower and upper bound calculations. We then study the variation of these bounds with respect to perturbations of the gas network availability process. The results of this work can be used in several ways: to study the investment return in dual-firing capabilities and oil storage capacity; to value gas provision contracts with low reliability; to value the benefits of gas network reliability improvements to power generation operators.

1.1 Context and Related Work

Natural gas fired electric power generators make an important part of the U.S. electricity generation mix [6]. However, gas-fired generators can be affected by gas supply disruptions. For instance, in some severe weather situations such as cold weather, high demand for gas and high demand for electricity put the gas system under stress, which can go as far as interrupting pipeline gas supply and force gas-fired power generation outages. Power generation outages have a significant effect on work and life and can disrupt the functioning of many critical sites such as hospitals, airports, and manufacturing facilities.

To mitigate the risk of gas access disruptions , power generator units can be equipped with dual-firing capabilities, to be able to use another fuel when gas is unavailable [30]. Dual-firing capabilities bring resiliency to the electric system, assuming that the alternate fuel can be used during temporary interruptions of gas supply [23]. Typical alternate fuels are petroleum and coal. Both can be stored onsite to mitigate the dependence upon their own delivery infrastructure (storing natural gas onsite would necessitate cryogenic facilities which may be expensive to operate and difficult to stabilize during a power outage). In this paper, we focus on oil as the alternate fuel and assume the existence of onsite oil storage capacity. Power generation from petroleum is marginal in the U.S. due to high costs and emissions. Nevertheless oil-fired generators are used during peak load times and their high cost is a factor contributing to driving up the spot price of electricity.

The literature on the valuation of gas-fired electric power generation is extensive, especially in the applied energy finance literature. Gas generators are often viewed as a collection of call options on the "spark spread", which is the difference between the price of 1 MWh of electric energy versus the price of gas multiplied by the quantity needed to produce the electric MWh. See [3, 4, 21]. When the spark spread is positive, gas is converted into electric power. Flexibility in gas procurement contracts is often valued as a swing option, which gives the holder the obligation to withdraw a prescribed minimal cumulated amount of gas and the option to withdraw a maximal cumulated amount, over a contractual time window, subject to minimal and maximal amounts at each exercise. At each exercise, the holder receives the difference between the gas spot price and the contractual price. See [13, 24].

Several mathematical models and methods have been proposed to study the dependency between power generation and the gas network. For instance, [22] employ a two-stage nonlinear optimization model for coordinating operations of the power system and natural gas network. Koeppel and Andersson [15] use simulation to analyze the extent to which the electrical network can be affected by the gas network. Li et al. [18] study the influence of interdependency of natural gas network and electricity infrastructure on power system security. Fleten and Näsäkkälä [9] consider generation expansion problems under stochastic electricity and natural gas prices. However, few structural results have been developed.

The purpose of this paper is to assess the benefit of having oil as an alternative fuel to natural gas to operate a combustion turbine. To do this, we use stylized models to describe uncertainties. We make the assumption that the gas network access is a Markov process that can be described by a finite Markov chain. We describe the price of electricity, price of oil, and price of gas as a mean-reverting process (in log scale). These are reasonable approximations. For instance, [2, 5, 7, 14, 20] use mean-reverting processes to describe electricity spot prices, [25] introduced a mean-reverting model to describe commodity price dynamics, [10] tested the term structure of futures prices of oil and natural gas from 1994 to 1999, and found mean reversion in these two commodities.

1.2 Contributions and Organization

The contributions of this paper can be summarized as follows. We incorporate the uncertainty of the gas network into a dual-firing power generation problem. We formulate the control problem as a Markov decision process where the electricity price, gas price, and oil price follow a mean-reverting process (in log scale) and the availability of gas is described as a Markov chain. We analyze the structure of an equivalent multistage stochastic optimization formulation. We develop lower and upper bounds on the value function. We carry out the perturbation analysis of the lower and upper bounds over the parameters of the gas availability process.

Technical contributions. Technically, our paper works as follows. We first formulate a Markov decision process with continuous unbounded state variables and discrete bounded decision variables, and then reformulate it as a multistage mixed-integer stochastic optimization problem. The first result is Proposition 1 which shows that the problem is in fact totally unimodular (TU), and can thus be approached using tools from convex stochastic optimization [28]. Totally, unimodular multistage stochastic mixed-integer optimization is discussed in [29]. The authors use finite random variables and rely on an extensive-form formulation. Unlike [29], we use a value function to shield us from the exponential growth of the problem dimension with the number of stages. This also allows us to handle continuous random variables. Our formulation circumvents difficulties that arise with different sets of assumptions (see, e.g., [12, 16, 17]) because in our problem, the integrality of decisions can be propagated forward in time, the stochastic process component that affects the feasible sets is integral, and our representation of the expected value function at the next state provably preserves the TU structure.

The second result, stated in Propositions 2 and 3, describes a lower bound and its sensitivity to perturbations of the problem data. The method consists in using a suboptimal solution that decouples the problem into independent parts, for which closed-form computations with continuous random variables are possible. The third result, stated in Proposition 4, describes the sensitivity of an upper bound obtained by assuming perfect foresight of the continuous random variables, while keeping the uncertainty for the discrete random variables. This technique works because the subproblems with perfect foresight are finite Markov decision processes, which can be solved exactly by dynamic programming techniques. The last result, Proposition 5, shows that the lower and upper bounds give the exact solution to a degenerate version of the problem.

Organization. The remainder of the paper is organized as follows. Section 2 describes the system model in the framework of Markov decision processes. Section 3 furnishes an equivalent formulation in the framework of multistage stochastic optimization, from which structural results are obtained. Section 4 furnishes a lower bound on the value function and studies its sensitivity to perturbations of the gas availability process. Section 5 furnishes an upper bound and studies its sensitivity to perturbations of the gas availability process. Section 6 illustrates the results numerically, and Sect. 7 concludes.

2 Model Description

This section describes our gas-electric-oil system as a Markov decision process in discrete time over a finite horizon.

State variables. The state at time $t = 0, 1, \ldots, T$ is denoted S_t. The state is described by five state variables:

$$S_t = (l_t, p_t^e, p_t^g, p_t^o, b_t). \tag{1}$$

The variable l_t is the storage level of oil in the oil tank at the beginning of period t, p_t^e is the price of electricity, p_t^g is the price of gas, p_t^o is the price of oil, and b_t indicates whether the gas network is under disruption and unavailable ($b_t = 0$) or functions normally ($b_t = 1$). The storage level l_t is limited by K, which is the capacity of the oil tank. The prices are positive. The variables $\log(p_t^e)$, $\log(p_t^g)$, $\log(p_t^o)$ are called the log-prices.

Decisions variables. The decision at time $t = 0, \ldots, T-1$ is denoted A_t. The decision is described by the following decision variables:

$$A_t = (u_t^g, u_t^o, q_t) \tag{2}$$

The variables u_t^g, u_t^o are 0–1 indicators determining the fuel utilization during the current period. If $u_t^g = 1$, the unit converts gas to power. If $u_t^o = 1$, the unit converts oil to power. Those are mutually exclusive: $u_t^g + u_t^o \in \{0, 1\}$. The variable q_t is the quantity of oil that is ordered at the beginning period t. This quantity is delivered to the oil tank at the end of period t and is ready to be consumed at the beginning of the next period. The decision u_t^g is constrained by the availability of the gas network:

$$u_t^g \leq b_t. \tag{3}$$

The decisions u_t^o and q_t are subject to oil tank capacity constraints:

$$\begin{array}{rl} l_t + q_t \leq K & \text{if} \quad u_t^o = 0, \\ l_t - O_c \geq 0 \quad \text{and} \quad l_t - O_c + q_t \leq K & \text{if} \quad u_t^o = 1. \end{array} \tag{4}$$

The first constraint in (4) applies when oil is not consumed during the current period. It says that the oil level postdelivery cannot exceed the oil storage capacity K. The second constraint applies when oil is used as fuel during the current period. It says that the oil level must be enough to satisfy the consumption, and then that the oil level post-consumption and postdelivery cannot exceed the capacity. The quantity O_c describes the quantity oil consumed to produce a quantity C of electricity (more on this in the definition of the reward function below).

We assume that the starting level l_0 and the quantity q_t are restricted to nonnegative multiples of O_c, and that K is a multiple of O_c (by rounding K down if necessary). This implies that $l_t \in \{0, O_c, 2O_c, \ldots, K\}$ and $q_t \in \{0, O_c, 2O_c, \ldots, K, K + O_c\}$, that is, the decision space is finite. In this context, we also assume that $K \geq O_c$, otherwise oil will never be used.

Rewards. The reward $R_t = R(S_t, A_t)$ for $t = 0, \ldots, T-1$ gives the expected reward of being in state S_t and making decision A_t. The reward function is described as follows:

$$\begin{array}{lll} R(S_t, A_t) = -p_t^o q_t & \text{if} & u_t^g = 0 \text{ and } u_t^o = 0, \\ R(S_t, A_t) = C p_t^e - G_c p_t^g - p_t^o q_t & \text{if} & u_t^g = 1, \\ R(S_t, A_t) = C p_t^e - p_t^o q_t & \text{if} & u_t^o = 1. \end{array} \quad (5)$$

The term $-p_t^o q_t$ is the cost of ordering a quantity q_t of oil at the spot price of oil $p_t^o = \exp(\log(p_t^o))$. The term $C p_t^e$ is the revenue from producing a quantity C of electricity at the spot price of electricity p_t^e. We assume that the generating unit is all-or-nothing and that the time the unit runs per period is one hour. In this case, C is also the power capacity of the turbine. The term $G_c p_t^g$ is the cost of the gas needed to produce the quantity C of electricity. The cost depends on the quantity G_c of gas consumed to produce C and on the spot price of natural gas p_t^g. No term involves the price of oil when electricity is produced from oil because oil is withdrawn from the oil tank at that time.

We also define a terminal reward $R_T(S_T)$ at the terminal time $t = T$,

$$R_T(S_T) = p_T^o l_T. \quad (6)$$

The term $p_T^o l_T$ represents the revenue from the sale of the oil that remains in the tank at the end of the horizon, valued at the spot price of oil.

State Transitions. The log-prices follow a mean-reverting process in discrete time. Let Δ denote the period duration. Let κ^e, κ^g, κ^o be positive parameters for the mean reversion rates. We assume that Δ is made small enough such that the κ parameters are in $(0, 1/\Delta)$. Let $\log(\zeta^e)$, $\log(\zeta^g)$, $\log(\zeta^o)$ be parameters for the mean levels. Let $\sigma^e, \sigma^g, \sigma^o$ be positive parameters for the volatilities. Let $W_{t+1} = (W_{t+1}^e, W_{t+1}^g, W_{t+1}^o)$ be a Gaussian multivariate random vector with zero mean, unit variances, and correlation matrix Σ. The vectors W_{t+1} at different time steps are mutually independent. We define

$$\begin{array}{l} \log p_{t+1}^e - \log p_t^e = \kappa^e(\log(\zeta^e) - \log p_t^e)\Delta + \sigma^e \sqrt{\Delta} W_{t+1}^e \\ \log p_{t+1}^g - \log p_t^g = \kappa^g(\log(\zeta^g) - \log p_t^g)\Delta + \sigma^g \sqrt{\Delta} W_{t+1}^g \\ \log p_{t+1}^o - \log p_t^o = \kappa^o(\log(\zeta^o) - \log p_t^o)\Delta + \sigma^o \sqrt{\Delta} W_{t+1}^o \,. \end{array} \quad (7)$$

The gas network state b_t follows a Markov chain with two states labeled 0,1. The transition matrix P of the gas network Markov chain is given by

$$P = \begin{bmatrix} p_{00} & p_{01} \\ p_{10} & p_{11} \end{bmatrix}, \qquad \begin{array}{l} p_{01} = 1 - p_{00} \in (0, 1) \\ p_{10} = 1 - p_{11} \in (0, 1). \end{array} \quad (8)$$

The oil level state variable l_t evolves as follows:

$$\begin{aligned} l_{t+1} &= l_t + q_t \quad &\text{if} \quad u_t^o = 0, \\ l_{t+1} &= l_t - O_c + q_t \quad &\text{if} \quad u_t^o = 1. \end{aligned} \tag{9}$$

Assuming $l_0 \in [0, K]$, the constraints guarantee that l_{t+1} remains in $[0, K]$.

Objective. We maximize the expected discounted cumulated reward over the finite horizon. For this objective, the minimal standard set of policies that contains an optimal policy is the set of nonstationary deterministic Markov policies. Let π be an admissible policy from that set, i.e., the decisions also satisfy the constraints. Viewing S_t, A_t as stochastic processes, let $\mathbb{P}_{s_0}^{\pi}$ and $\mathbb{E}_{s_0}^{\pi}$ denote the probability measure and expectation operator induced by the stochastic state transitions, the choice of A_t according to policy π, and the initial state s_0. Let $\gamma \in (0, 1]$ be the discount factor. The objective is then written as

$$V_0(s_0) = \max_{\pi} \mathbb{E}_{s_0}^{\pi} \left\{ \sum_{t=0}^{T-1} \gamma^t R(S_t, A_t) + \gamma^T R_T(S_T) \right\}. \tag{10}$$

Optimality Conditions. Following Bellman's optimality principle, the optimality conditions can be expressed recursively in terms of the value functions V_t:

$$\begin{aligned} V_T(s) &= R_T(s), \\ V_t(s) &= \sup_a \{R(s, a) + \gamma \mathbb{E}\{V_{t+1}(S_{t+1}) \mid S_t = s, A_t = a\}\} \end{aligned} \tag{11}$$

for $t = T - 1, \ldots, 1, 0$.

3 Equivalent Multistage Stochastic Optimization Formulation

Let $q_t^o = q_t / O_c$, $l_t^o = l_t / O_c$, $K^o = K / O_c$ be quantities rescaled to be integer-valued. Suppose momentarily that p_t^e, p_t^g, p_t^o, b_t are all fixed, and given l_0^o, consider the problem

$$\begin{aligned} \text{maximize} \quad & \sum_{t=0}^{T-1} \gamma^t [C p_t^e (u_t^g + u_t^o) - G_c p_t^g u_t^g - O_c p_t^o q_t^o] + \gamma^T p_T^o O_c l_T^o \\ \text{subject to} \quad & u_t^g \le b_t, \quad u_t^o \le l_t^o, \quad u_t^g + u_t^o \le 1, \\ & l_{t+1}^o = l_t^o - u_t^o + q_t^o, \quad l_{t+1}^o \le K^o, \\ & (u_t^g, u_t^o, q_t^o, l_{t+1}^o) \in \mathbb{Z}_+^4 \qquad \text{for } t = 0, \ldots, T - 1, \end{aligned} \tag{12}$$

where $\mathbb{Z}_+$ is the set of nonnegative integers. It can be checked that this formulation is equivalent to the problem stated in Sect. 2 posed over a fixed realization of the prices and gas network states. By eliminating $q_t^o = l_{t+1}^o + u_t^o - l_t^o$, but enforcing its nonnegativity, we get the following description of the feasible set,

$$\mathscr{U}_t(l_t^o; b_t) := \{(u_t^g, u_t^o, l_{t+1}^o) \in \mathbb{Z}_+^3 : \; u_t^g \le b_t, \; u_t^o \le l_t^o, \quad u_t^g + u_t^o \le 1, \\ l_{t+1}^o + u_t^o \ge l_t^o, \quad l_{t+1}^o \le K^o \}. \tag{13}$$

Letting $(p_t^e, p_t^g, p_t^o, b_t)$ be a random process again and adapting the decisions to the generated filtration leads to a multistage stochastic programming formulation of the original problem. We state it below in nested form, using the notation V_t for the value functions, since except for the change of variables, they coincide with the value functions V_t in (11):

$$V_T(l_T^o; p_T^e, p_T^g, p_T^o, b_T) = p_T^o O_c l_T^o,$$
$$V_t(l_t^o; p_t^e, p_t^g, p_t^o, b_t) = \max_{(u_t^g, u_t^o, l_{t+1}^o) \in \mathscr{U}_t(l_t^o; b_t)} \{(Cp_t^e - G_c p_t^g)u_t^g + (Cp_t^e - O_c p_t^o)u_t^o \\ - O_c p_t^o(l_{t+1}^o - l_t^o) + \gamma \, \mathscr{V}_t(l_{t+1}^o; p_t^e, p_t^g, p_t^o, b_t)\}, \tag{14}$$

$$\mathscr{V}_t(l_{t+1}^o; p_t^e, p_t^g, p_t^o, b_t) := \mathbb{E}\{V_{t+1}(l_{t+1}^o; p_{t+1}^e, p_{t+1}^g, p_{t+1}^o, b_{t+1}) \mid p_t^e, p_t^g, p_t^o, b_t\}. \tag{15}$$

Proposition 1 *Let l_t^o be integer, but suppose that the integrality constraints in $\mathscr{U}_t(l_t^o;\, b_t)$ are relaxed. Suppose that $\mathscr{V}_t(l_{t+1}^o; p_t^e, p_t^g, p_t^o, b_t)$ is extended to a continuous domain for l_{t+1}^o by piecewise-linear interpolation. Then, it holds that the integrality of the optimal decisions $(u_t^g, u_t^o, l_{t+1}^o)$ is preserved, and that there is no gain or loss in optimality. Furthermore, the piecewise-linear interpolant for V_t is actually concave in l_t^o, which implies that the value function V_t has nonincreasing differences in l_t^o, for almost all $(p_t^e, p_t^g, p_t^o, b_t)$.*

Having nonincreasing differences in l_t^o means that as the tank fills up, the extra value from having an extra unit of stored oil diminishes or at best stays the same.

Proof. Suppose that in (14), the function $\mathscr{V}_t(l_{t+1}^o; p_t^e, p_t^g, p_t^o, b_t)$ is replaced by a continuous piecewise-linear interpolant $\phi_t(l_{t+1}^o)$, chosen to coincide with $\mathscr{V}_t(l_{t+1}^o; p_t^e, p_t^g, p_t^o, b_t)$ at integer l_{t+1}^o. Note that $\mathscr{V}_t$ and thus ϕ_t are nondecreasing in l_{t+1}^o, and actually increasing, since increasing l_{t+1}^o relaxes the oil inventory constraints, while the increment can be used either to offset the cost of an optimal oil purchase, or otherwise to increase the terminal reward.

Suppose inductively that ϕ_t is also concave in l_{t+1}^o. Concavity holds at $t = T - 1$, since V_T for all p_T^o and thus $\mathscr{V}_{T-1}$ are affine in l_T^o. Then, $\phi_t(l_{t+1}^o)$ admits the following representation (with state-dependent coefficients):

$$\phi_t(l_{t+1}^o) = \max_{y \in \mathscr{Y}(l_{t+1}^o)} \{c_0 + \sum_{i=1}^{K^o} c_i y_i\},$$
$$\mathscr{Y}(l_{t+1}^o) = \{y \in \mathbb{R}^{K^o} : 0 \le y_i \le 1, \; \sum_{i=1}^{K^o} y_i = l_{t+1}^o\},$$
$$c_0 := \mathscr{V}_t(0; p_t^e, p_t^g, p_t^o, b_t),$$
$$c_i := \mathscr{V}_t(i; p_t^e, p_t^g, p_t^o, b_t) - \mathscr{V}_t(i-1; p_t^e, p_t^g, p_t^o, b_t), \tag{16}$$

where $c_1 \geq c_2 \geq \cdots \geq c_{K^o}$ by concavity of $\mathscr{V}_t$. We make the change of variable $z_i = \sum_{j=1}^{i} y_j$ and set by convention $z_0 = 0$. Therefore,

$$y_i = z_i - z_{i-1},$$
$$\sum_{i=1}^{K^o} c_i y_i = \sum_{i=1}^{K^o} c_i (z_i - z_{i-1}) = c_{K^o} z_{K^o} + \sum_{i=1}^{K^o-1} (c_i - c_{i+1}) z_i.$$

Setting $h_i = c_{i+1} - c_i$, we get

$$\phi_t(l_{t+1}^o) = \max_{z \in \mathscr{Z}(l_{t+1}^o)} \{c_0 + c_{K^o} l_{t+1}^o - \sum_{i=1}^{K^o-1} h_i z_i\} \tag{17}$$
$$\mathscr{Z}(l_{t+1}^o) = \{z \in \mathbb{R}^{K^o} : 0 \leq z_i - z_{i-1} \leq 1,\ z_{K^o} = l_{t+1}^o\} \quad \text{(with } z_0 = 0\text{)}$$
$$h_i := \mathscr{V}_t(i-1; p_t^e, p_t^g, p_t^o, b_t) - 2\mathscr{V}_t(i; p_t^e, p_t^g, p_t^o, b_t) + \mathscr{V}_t(i+1; p_t^e, p_t^g, p_t^o, b_t).$$

Going back to (14), we make a last change of variable: we introduce $l_t^u = l_t^o - u_t^o$ (oil inventory at the end of the period just before replenishment). Thus, $u_t^o = l_t^o - l_t^u$ where l_t^o is fixed. Finally, we relax the integrality requirements of the set $\mathscr{U}_t(l_t^o, b_t)$. The input l_t^o is still assumed to be integer. This leads to the following continuous relaxation of the maximization problem in (14):

$$\begin{aligned}
\text{maximize}\quad & (Cp_t^e - G_c p_t^g)u_t^g - (Cp_t^e - O_c p_t^o)l_t^u + Cp_t^e l_t^o - O_c p_t^o l_{t+1}^o \\
& + \gamma \left[c_0 + c_{K^o} l_{t+1}^o - \sum_{i=1}^{K^o-1} h_i z_i \right] \\
\text{subject to}\quad & 0 \leq u_t^g \leq b_t, \quad 0 \leq l_t^u \leq l_t^o, \\
& l_t^u - u_t^g \geq l_t^o - 1, \quad l_{t+1}^o - l_t^u \geq 0, \quad 0 \leq l_{t+1}^o \leq K^o, \\
& 0 \leq l_{t+1}^o - z_{K^o-1} \leq 1, \quad 0 \leq z_1 \leq 1, \\
& 0 \leq z_i - z_{i-1} \leq 1 \quad \text{for } i = 2, \ldots, K^o - 1.
\end{aligned} \tag{18}$$

The constraints have the form $\mathbf{A}x \leq \mathbf{b}$. Recall that the vertex optimal solutions to a linear problem are integral if and only if $\mathbf{b}$ is integral and $\mathbf{A}$ is totally unimodular. A sufficient condition for $\mathbf{A}$ to be totally unimodular is that its elements are in $\{0, -1, +1\}$, each row has at most 2 nonzero elements, and if a row has two nonzero elements they must have opposite sign. Those conditions are verified with (18). The right-hand side $\mathbf{b}$ is integer since l_t^o is fixed and integer. Hence, the optimal vertex solutions to (18) are integer, and the relaxation is tight. Reverting back to the original variables, those are automatically integer as well. We have thus established that if l_t^0 is integer and the linear interpolant of $\mathscr{V}_t(\cdot; p_t^e, p_t^g, p_t^o, b_t)$ is concave, then the set $\mathscr{U}(l_t^o, b_t)$ can be replaced by its convex relaxation.

It remains to complete the induction argument. We start with the concavity of the linear interpolant. Let $x = (u_t^g, l_t^u, l_{t+1}^o, z_1, \ldots, z_{K^o-1})$, and let $\mathscr{X}(l_t^o, b_t)$ denote the feasible set for x in (18). Let $\varphi^*(l_t^o)$ denote the optimal value of (18) from the maximization over $x \in \mathscr{X}(l_t^o, b_t)$. Now, let $\mathscr{C} = \{(l_t^o, x) : x \in \mathscr{X}(l_t^o, b_t),\ 0 \leq l_t^o \leq K^o\}$. As a polytope, $\mathscr{C}$ is convex. The objective function of (18) is linear in (l_t^o, x) and thus jointly concave on $\mathscr{C}$, see, e.g., [8]. From these conditions, it follows that $\varphi^*(l_t^o)$ is concave in l_t^o. On the discrete values of l_t^o, we have $\varphi^*(l_t^o) = V_t(l_t^o; p_t^e, p_t^g, p_t^o, b_t)$. Taking the expectation over $(p_t^e, p_t^g, p_t^o, b_t)$ preserves the concavity properties in l_t^o, thus $\phi_{t-1}(l_t^o) = \mathbb{E}\{\varphi^*(l_t^o) | p_{t-1}^e, p_{t-1}^g, p_{t-1}^o, b_{t-1}\}$ is concave in l_t^o. Note that on the discrete values of l_t^o, $\phi_{t-1}(l_t^o) = \mathscr{V}_{t-1}(l_t^o; p_{t-1}^e, p_{t-1}^g, p_{t-1}^o, b_{t-1})$, so we have shown that the linear interpolant of $\mathscr{V}_{t-1}$ is concave in l_t^o, as required.

Finally, if l_t^o is integer, it has also been shown that l_{t+1}^o is integer almost surely. Thus, the value functions $V_t(l_t^o; p_t^e, p_t^g, p_t^o, b_t)$ for integer l_t^o only query the functions $V_{t+1}(l_{t+1}^o; p_{t+1}^e, p_{t+1}^g, p_{t+1}^o, b_{t+1})$ at integer l_{t+1}^o. This ensures one can always replace $\mathscr{V}_{t+1}$ by a piecewise continuous function with breakpoints at integer l_{t+1}^o without loss of optimality. □

An interesting technical detail in the proof is the description (17) of the expected value function at the next state. Usually, polyhedral approximations of concave (respectively, convex) value functions are described as the minimum (respectively, maximum) of linear functions, which can be converted to a family of inequality constraints (linear cuts), e.g., as in stochastic dual dynamic programming (SDDP) methods [27]. The description (17) is based instead on a maximum, while continuing to use primal variables. This representation turns out to be instrumental for establishing by (18) the integrality of the solution of the convex relaxation.

When referring back to (18), note that the coefficients c_0, c_{K^o} and h_i depend on p_t^e, p_t^g, p_t^o, b_t. Proposition 1 addresses concavity properties of the value function with respect to l_t^o, but does not address the structure of the value function with respect to the price state variables.

Finally, we note that the mathematical result of Proposition 1 has a technological interpretation. The ability to *simultaneously* use different fuels is called *co-firing*. Co-firing is technologically more challenging than dual-firing. Mathematically, co-firing removes the integrality constraint on the variables u_t^e, u_t^o. The result of Proposition 1 shows that co-firing (of oil and gas) has *zero value* over dual-firing. This is of course under the assumption that the unit can operate under any single of the two fuels.

4 Lower Bounds

Lower bounds on the value function (11) can be obtained by calculating the value of a given policy. One possibility that leads to closed-form calculations is to run on gas when gas is available ($u_t^g = 1$ if $b_t = 1$ and $Cp_t^e - G_c p_t^g > 0$), and run on oil when gas is unavailable. To be able to run on oil without tracking the storage level l_t, we can select a suboptimal oil replenishment policy. For instance, we replenish with $q_t = O_c$ each time oil is used. In this case, it can make sense to use oil when $b_t = 0$ and $Cp_t^e - O_c p_t^o > 0$, having the guarantee that $l_t \geq O_c$.

The value function is then bounded by the sum of two independent value functions:

$$V(s_0) \geq V^{\text{gas}}(s_0) + V^{\text{oil}}(s_0), \tag{19}$$

where $V^{\text{gas}}(s_0)$ is the value of converting gas to power, which depends on the electricity gas spread ("spark spread" [3]) and on the gas network availability probabilities, and $V^{\text{oil}}(s_0)$ is the value of converting oil to power, which depends on the electricity oil spread and on the gas network unavailability probabilities.

The two value functions can be summed up in the right-hand side of (19) because their contributions to the expected reward at time t conditionally to b_t are mutually exclusive, the suboptimal fuel utilization policy does not compare the fuel prices, and the two policies relative to V^{gas} and V^{oil} affect noninteracting components of the state. Actually, in the special case where the recourse to oil is nonexistent (for instance if $K = 0$), $V^{\text{gas}}(s_0)$ would coincide with the exact optimal value $V(s_0)$ of the problem.

To calculate the bounds, we first write (7) in matrix form. Let $\xi_t = [\log p_t^e \ \log p_t^g \ \log p_t^o]^\top$. We write $z_{t+1} \sim \mathcal{N}(\mu_z, \Sigma_z)$ to indicate that z_{t+1} follows a multivariate Gaussian of mean μ_z and covariance matrix Σ_z. We have

$$\begin{aligned}
&\xi_{t+1} = D\xi_t + z_{t+1}, \qquad D = \begin{bmatrix} (1-\kappa^e\Delta) & 0 & 0 \\ 0 & (1-\kappa^g\Delta) & 0 \\ 0 & 0 & (1-\kappa^o\Delta) \end{bmatrix}, \\
&z_{t+1} \sim \mathcal{N}(\mu_z, \Sigma_z), \\
&\mu_z = \Delta \begin{bmatrix} \kappa^e \log(\zeta^e) \\ \kappa^g \log(\zeta^g) \\ \kappa^o \log(\zeta^o) \end{bmatrix}, \quad \Sigma_z = \Delta \begin{bmatrix} (\sigma^e)^2 & \sigma^e\sigma^g\Sigma_{12} & \sigma^e\sigma^o\Sigma_{13} \\ \sigma^e\sigma^g\Sigma_{12} & (\sigma^g)^2 & \sigma^g\sigma^o\Sigma_{23} \\ \sigma^e\sigma^o\Sigma_{13} & \sigma^g\sigma^o\Sigma_{23} & (\sigma^o)^2 \end{bmatrix}.
\end{aligned} \tag{20}$$

From there, ξ_t given ξ_0 follows $\mathcal{N}(\bar{x}_t, X_t)$ where $\bar{x}_t = \begin{bmatrix} \bar{x}_{t,e} \\ \bar{x}_{t,g} \\ \bar{x}_{t,o} \end{bmatrix}$, $X_t = \begin{bmatrix} X_{t,ee} & X_{t,eg} & X_{t,eo} \\ X_{t,eg} & X_{t,gg} & X_{t,go} \\ X_{t,eo} & X_{t,go} & X_{t,oo} \end{bmatrix}$ are defined by the recursion

$$\bar{x}_{t+1} = D\bar{x}_t + \mu_z, \quad X_{t+1} = DX_tD^\top + \Sigma_z, \quad \text{with} \quad \bar{x}_0 = \xi_0, \quad X_0 = 0. \tag{21}$$

One has $\bar{x}_t = \sum_{k=0}^{t-1} D^k\mu_z + D^t\xi_0$ and $X_t = \sum_{k=0}^{t-1} D^k\Sigma_z D^k$ for $t \geq 1$.

Let

$$h_t^{eg} = Cp_t^e - G_c p_t^g. \tag{22}$$

To get the best lower bound, the optimal decision given $b_t = 1$ is $u_t^g = 1$ if $h_t^{eg} > 0$, and $u_t^g = u_t^o = 0$ otherwise. The probability that $b_t = 1$ is described by the second element of the row vector

$$[\mathbb{P}(b_t = 0), \mathbb{P}(b_t = 1)] := g_t = g_0 P^t := [g_{t0}, g_{t1}]. \tag{23}$$

It follows that

$$V^{\text{gas}}(s_0) = g_{01}[h_0^{eg}]^+ + \sum_{t=1}^{T-1} \gamma^t g_{t1} \mathbb{E}\{[h_t^{eg}]^+\}, \tag{24}$$

$$[\log(Cp_t^e), \log(G_c p_t^g)]^\top \sim \mathcal{N}(\bar{y}_t, Y_t),$$

$$\bar{y}_t = \begin{bmatrix} \log C + \bar{x}_{t,e} \\ \log G_c + \bar{x}_{t,g} \end{bmatrix}, \quad Y_t = \begin{bmatrix} X_{t,ee} & X_{t,eg} \\ X_{t,eg} & X_{t,gg} \end{bmatrix}.$$

It remains to evaluate $\mathbb{E}\{[h_t^{eg}]^+\}$. Let $\Phi(\cdot)$ be the cumulative distribution function (cdf) of the standard normal distribution. By techniques similar to those used to establish Margrabe's formula [21], or by observing that $\mathbb{E}\{[h_t^{eg}]^+\} = \mathbb{E}\{Cp_t^e\} - \mathbb{E}\{\min\{Cp_t^e, G_c p_t^g\}\}$ and then using [19], one finds

$$\begin{aligned} \mathbb{E}\{[h_t^{eg}]^+\} &= C \exp(\bar{x}_{t,e} + \tfrac{1}{2} X_{t,ee}) \Phi\left(\frac{(\log(C/G_c) + \bar{x}_{t,e} - \bar{x}_{t,g}) + X_{t,ee} - X_{t,eg}}{\sqrt{X_{t,ee} + X_{t,gg} - 2X_{t,eg}}} \right) \\ &\quad - G_c \exp(\bar{x}_{t,g} + \tfrac{1}{2} X_{t,gg}) \Phi\left(\frac{(\log(C/G_c) + \bar{x}_{t,e} - \bar{x}_{t,g}) - X_{t,gg} + X_{t,eg}}{\sqrt{X_{t,ee} + X_{t,gg} - 2X_{t,eg}}} \right). \end{aligned} \tag{25}$$

For consistency, note that $\mathbb{E}\{[h_0^{eg}]^+\} = h_0^{eg}$ by letting X_0 tend to 0.

The calculations for $V^{\text{oil}}(s_0)$ are similar, except that $V^{\text{oil}}(s_0)$ is based on two components, $V^{\text{oil},1}(s_0)$ and $V^{\text{oil},2}(s_0)$. The first component ensures we can always run on oil at times $t \geq 1$. Recall that $K \geq O_c$ by assumption. If $l_0 < O_c$ at time 0, we order O_c. It will be left to the second component to order additional oil if we consume oil at time 0, and to manage the balance of orders with consumption onwards. Then, $l_t = \max\{l_0, O_c\}$ for all $t \geq 1$. At time T, the oil level l_T is liquidated via the terminal reward. Overall we have

$$V^{\text{oil},1}(s_0) = 1_{\{l_0 < O_c\}}(-O_c p_0^o) + \gamma^T \max\{l_0, O_c\} \mathbb{E}\{p_T^o\}, \tag{26}$$

$$\mathbb{E}\{p_T^o\} = \exp(\bar{x}_{T,o} + X_{T,oo}/2). \tag{27}$$

The second component is the return from the conversion from oil to power. Recall that $g_{t0} = 1 - g_{t1}$ is the probability that the gas network is unavailable at time t. Let $h_t^{eo} = Cp_t^e - O_c p_t^o$ be the electric-oil spread, noting that $-O_c p_t^o$ is actually the cost of replenishing for the next time oil is used. Setting time 0 apart to check the feasibility of the oil-electric conversion at the initial time, we get

$$V^{\text{oil},2}(s_0) = g_{00} 1_{\{l_0 \geq O_c\}} [h_0^{eo}]^+ + \sum_{t=1}^{T-1} \gamma^t g_{t0} \mathbb{E}\{[h_t^{eo}]^+\}, \tag{28}$$

$$\begin{aligned}\mathbb{E}\{[h_t^{eo}]^+\} = C \exp(\bar{x}_{t,e} + \tfrac{1}{2} X_{t,ee}) \Phi \left(\frac{(\log(C/O_c) + \bar{x}_{t,e} - \bar{x}_{t,o}) + X_{t,ee} - X_{t,eo}}{\sqrt{X_{t,ee} + X_{t,oo} - 2X_{t,eo}}} \right) \\ - O_c \exp(\bar{x}_{t,o} + \tfrac{1}{2} X_{t,oo}) \Phi \left(\frac{(\log(C/O_c) + \bar{x}_{t,e} - \bar{x}_{t,o}) - X_{t,oo} + X_{t,eo}}{\sqrt{X_{t,ee} + X_{t,oo} - 2X_{t,eo}}} \right).\end{aligned} \tag{29}$$

The quality of $V^{\text{oil},1} + V^{\text{oil},2}$ is negatively impacted by the suboptimality of the oil replenishment strategy. Trying to improve over the replenishment strategy would lead to a formulation similar to the original problem, we set out to solve, with one fewer state variable (the price of gas). We can nevertheless improve the quality of the bound by considering other suboptimal policies that are easy to value. For instance, the expected return of doing nothing with oil is $\gamma^T l_0 \mathbb{E}\{p_T^o\} \geq 0$, from the terminal reward. In light of this remark, we can use

$$V^{\text{oil}}(s_0) = \max\{V^{\text{oil},1}(s_0) + V^{\text{oil},2}(s_0), \gamma^T l_0 \mathbb{E}\{p_T^o\}\}. \tag{30}$$

To summarize, we have shown that $V(s_0) \geq V^{LB}(s_0) := V^{\text{gas}}(s_0) + V^{\text{oil}}(s_0)$ with $V^{\text{gas}}(s_0)$ given by (24) and $V^{\text{oil}}(s_0)$ given by (30). These bounds offer a first look into the sensitivity of the problem value to the parameters describing the reliability of the gas network:

Proposition 2 *The lower bound $V^{LB}(s_0)$ varies linearly in the gas network availability probabilities* (23), *except for a kink at the point where $V^{oil,1} + V^{oil,2} = \gamma^T l_0 \mathbb{E}\{p_T^o\}$.*

Proof. Suppose that $V^{\text{oil}}(s_0) = V^{\text{oil},1}(s_0) + V^{\text{oil},2}(s_0) > \gamma^T l_0 \mathbb{E}\{p_T^o\}$, see (30), which means that using oil backup is justified economically. Then we have, from (24), (28) and $g_{t0} = 1 - g_{t1}$,

$$\frac{\partial V^{\text{LB}}(s_0)}{\partial g_{t1}} = \begin{cases} [h_0^{eg}]^+ - 1_{\{l_0 \geq O_c\}} [h_0^{eo}]^+ & \text{for } t = 0, \\ \gamma^t (\mathbb{E}\{[h_t^{eg}]^+\} - \mathbb{E}\{[h_t^{eo}]^+\}) & \text{for } t = 1, \ldots, T-1. \end{cases} \tag{31}$$

Alternatively, suppose that $V^{\text{oil}}(s_0) = \gamma^T l_0 \mathbb{E}\{p_T^o\} < V^{\text{oil},1}(s_0) + V^{\text{oil},2}(s_0)$. To know if oil backup is justified economically, we would need to consider a better oil replenishment strategy. With our current lower bound, we can only assert that

$$\frac{\partial V^{\text{LB}}(s_0)}{\partial g_{t1}} = \begin{cases} [h_0^{eg}]^+ & \text{for } t = 0, \\ \gamma^t \mathbb{E}\{[h_t^{eg}]^+\} & \text{for } t = 1, \ldots, T-1. \end{cases} \tag{32}$$

At the point where $V^{\text{oil},1}(s_0) + V^{\text{oil},2}(s_0) = \gamma^T l_0 \mathbb{E}\{p_T^o\}$, $V^{\text{LB}}(s_0)$ is in general not differentiable. □

The parameters of the gas network transition matrix P in (8) can be related to the probabilities g_{t1}. The dependence is nonlinear. To see this, consider the stationary probabilities of the gas network Markov chain, written g_∞ (row vector). Their derivatives can be described as $\dot{g}_\infty = g_\infty \dot{P}(I-P)^\#$, see [11], where $(I-P)^\# = (I-P+1g_\infty)^{-1} - 1g_\infty$, see [26], is the generalized group inverse of $I-P$, and the dot operation denotes differentiation with respect to a parameter of interest (which can be p_{00} or p_{11}). The steady-state probabilities are most sensitive to perturbations when p_{00} and p_{11} are both close to 1, i.e., rare but long gas network disruptions.

Proposition 3 below gives the detailed results. We consider perturbations such that the perturbed matrix is still a transition matrix. This means that, e.g., an increment of p_{00} is balanced by a decrement of p_{01} of equal magnitude, and the resulting probabilities should remain in (0, 1).

Proposition 3 *When $V^{oil}(s_0) \neq \gamma^T l_0 \mathbb{E}\{p_T^o\}$, the lower bound $V^{LB}(s_0)$ is differentiable in the parameters p_{00}, p_{11} of the gas network transition matrix. Its gradient is described by*

$$\frac{\partial V^{LB}(s_0)}{\partial p_{ii}} = \sum_{t=1}^{T-1} \gamma^t [g_0 \sum_{k=0}^{t-1} P^k \tfrac{\partial P}{\partial p_{ii}} P^{t-1-k}] \begin{bmatrix} \varepsilon^{oil}\mathbb{E}\{[h_t^{eo}]^+\} \\ \mathbb{E}\{[h_t^{eg}]^+\} \end{bmatrix} \quad \textit{for } i = 0, 1, \tag{33}$$

$$\varepsilon^{oil} = 1_{\{V^{oil}(s_0) > \gamma^T l_0 \mathbb{E}\{p_T^o\}\}}, \quad \frac{\partial P}{\partial p_{00}} := \begin{bmatrix} 1 & -1 \\ 0 & 0 \end{bmatrix}, \quad \frac{\partial P}{\partial p_{11}} := \begin{bmatrix} 0 & 0 \\ -1 & 1 \end{bmatrix}. \tag{34}$$

Proof. From $g_t = g_0 P^t$, we have $\dot{g}_t = g_0 \sum_{k=0}^{t-1} P^k \dot{P} P^{t-1-k}$. We introduce ε^{oil} to unify the results of Proposition 2. Then, the differentiation of $V^{LB}(s_0)$ with respect to p_{00}, p_{11} at a point where V^{LB} is differentiable ($V^{\text{oil},1}(s_0) + V^{\text{oil},2}(s_0) \neq \gamma^T l_0 \mathbb{E}\{p_T^o\}$) gives the expression of the proposition. □

5 Upper Bounds

Upper bounds on the value function (11) can be obtained by relaxing constraints and/or allowing decisions with foresight. Bounds based on perfect information are easy to write down but are often loose. One improvement is to assume perfect foresight of future prices, while keeping the gas network process b_t random. This is formalized as follows: we have

$$\begin{aligned} V(s_0) &= \max_\pi \mathbb{E}_{s_0}^\pi \{\sum_{t=0}^{T-1} \gamma^t R(S_t, A_t) + \gamma^T R_T(S_T)\} \\ &\leq V^{\text{UB}}(s_0) = \mathbb{E}_\xi \{V^\xi(s_0)\}, \end{aligned} \tag{35}$$

$$V^\xi(s_0) := \max_{\tilde{\pi}} \mathbb{E}_{\tilde{s}_0}^{\tilde{\pi}} \{\sum_{t=0}^{T-1} \gamma^t R_t^\xi(\tilde{S}_t, A_t) + \gamma^T R_T^\xi(\tilde{S}_T)\}. \tag{36}$$

The expectation $\mathbb{E}_\xi$ is over the price process, given s_0. Since ξ is treated as a fixed input of V^ξ, in the dynamic programming problem (36), ξ_t is removed from the state space. The reduced state is denoted $\tilde{S}_t = (l_t, b_t)$. The reduced state space is finite with cardinality $2(1 + K/O_c)$. The notation R_t^ξ emphasizes that the reward function is nonstationary, as it depends on ξ_t which is time-dependent. The reward is defined as $R_t^\xi(\tilde{S}_t, A_t) = R(S_t^\xi, A_t)$ where $S_t^\xi = (\xi_t, \tilde{S}_t) = (p_t^e, p_t^g, p_t^o, b_t, l_t)$. Similarly, $R_T^\xi(\tilde{S}_t) = R_T(S_t^\xi)$. The policy $\tilde{\pi}$, which is specific to ξ, maps reduced states to decisions A_t. The finite-horizon problem (36) can be solved exactly in T iterations by value iteration, using the auxiliary value functions $V_T^\xi(l_T, b_T) = l_T p_T^o$ and

$$V_t^\xi(l_t, b_t) = \max_{A_t}\{R_t^\xi(l_t, b_t, A_t) + \gamma[p_{b_t,0} V_{t+1}^\xi(\ell_{t+1}, 0) + p_{b_t,1} V_{t+1}^\xi(\ell_{t+1}, 1)]\} \tag{37}$$

for $t = T-1, \ldots, 0$, where we use the post-decision state $\ell_{t+1} = l_t - O_c u_t^o + q_t$, given (l_t, b_t) and $A_t = (u_t^g, u_t^o, q_t)$. One then sets $V^\xi(s_0) = V_0^\xi(l_0, b_0)$.

The expectation over ξ in (35) can be estimated by sample average approximation methods [28] and will produce a statistical upper bound. Typically, using a large number M of independent samples $\xi^{(m)}$, one calculates

$$\bar{V}_M = \tfrac{1}{M}\sum_{m=1}^{M} V^{\xi^{(m)}}(s_0), \quad \bar{\sigma}_M = \tfrac{1}{M}\sqrt{\sum_{m=1}^{M}[V^{\xi^{(m)}}(s_0) - \bar{V}_M]^2}, \tag{38}$$

and then adopt a statistical upper bound such as $\hat{V}^{\text{UB}}(s_0) = \bar{V}_M + 1.96\bar{\sigma}_M$ which holds with approximate confidence 97.5%.

Proposition 4 below describes the sensitivity of the upper bound (35) to perturbations of the gas network transition matrix.

Proposition 4 *Let $\ell_{t+1,j}$ denote the optimal ℓ_{t+1} when being in state (l_t, j) at time t, for $j = 0, 1$, as determined by (37). Define the matrix $\left[\frac{\partial P}{\partial p_{ii}}\right]$ as in (34).*

Define $\dfrac{\partial V_0^\xi(l_0, b_0)}{\partial p_{ii}}$ *recursively using* $\dfrac{\partial V_{T-1}^\xi}{\partial p_{ii}} = 0$ *and*

$$\frac{\partial V_t^\xi(l_t, j)}{\partial p_{ii}} = \gamma\left(\left[\frac{\partial P}{\partial p_{ii}}\right]\begin{bmatrix} V_{t+1}^\xi(\ell_{t+1,j}, 0) \\ V_{t+1}^\xi(\ell_{t+1,j}, 1)\end{bmatrix} + P\begin{bmatrix} \partial V_{t+1}^\xi(\ell_{t+1,j}, 0)/\partial p_{ii} \\ \partial V_{t+1}^\xi(\ell_{t+1,j}, 1)/\partial p_{ii}\end{bmatrix}\right)_j, \tag{39}$$

where $(\cdot)_j$ extracts the $(j+1)$th element. Then it holds that

$$\partial \mathbb{E}_\xi\{V^\xi(s_0)\}/\partial p_{ii} = \mathbb{E}_\xi\{\partial V_0^\xi(l_0, b_0)/\partial p_{ii}\}. \tag{40}$$

Proof. The maximization problem of (37) can be expressed in the form (14), except that the expectation in (15) is only over b_{t+1} given b_t. Thus, one can relax the integrality constraints and reason on the convex relaxation of the problem. The continuous relaxation of the feasible set $\mathscr{U}_{t+1}(l_t^o; b_t)$ is nonempty and compact. Recall that $l_t^o = l_t/O_c$. Slightly abusing notation, we write (37) in terms of l_t^o and l_{t+1}^o:

$$V_t^\xi(l_t^o, b_t) = \max_{(u_t^g, u_t^o, l_{t+1}^o) \in \mathscr{U}_{t+1}(l_t^o; b_t)} f_{l_t^o, b_t}^\xi(u_t^g, u_t^o, l_{t+1}^o)$$

$$f_{l_t^o, b_t}^\xi(u_t^g, u_t^o, l_{t+1}^o) := (Cp_t^e - G_c p_t^g)u_t^g + (Cp_t^e - O_c p_t^o)u_t^o - O_c p_t^o(l_{t+1}^o - l_t^o) \\ + \gamma[p_{b_t,0} V_{t+1}^\xi(l_{t+1}^o, 0) + p_{b_t,1} V_{t+1}^\xi(l_{t+1}^o, 1)].$$

At time T, $V_T^\xi(l_T^o, 0) = V_T^\xi(l_T^o, 1) = O_c p_T^o l_T^o$ is independent of b_T, p_{00}, and p_{11}. At time $T-1$, the objective is differentiable in p_{00} and p_{11}; using $p_{b_t,1} = 1 - p_{b_t,0}$ we have

$$\begin{aligned}
&\partial f_{l_{T-1}^o,0}^\xi / \partial p_{00} = \gamma[V_T^\xi(l_{T,0}^o, 0) - V_T^\xi(l_{T,0}^o, 1)] = 0, \\
&\partial f_{l_{T-1}^o,1}^\xi / \partial p_{00} = 0, \\
&\partial f_{l_{T-1}^o,0}^\xi / \partial p_{11} = 0, \\
&\partial f_{l_{T-1}^o,1}^\xi / \partial p_{11} = \gamma[-V_T^\xi(l_{T,1}^o, 0) + V_T^\xi(l_{T,1}^o, 1)] = 0.
\end{aligned}$$

Hence $\partial V_{T-1}^\xi / \partial p_{00} = \partial V_{t+1}^\xi / \partial p_{11} = 0$. At time $T-2$ we have

$$\begin{aligned}
&\partial f_{l_{T-2}^o,0}^\xi / \partial p_{00} = \gamma[V_{T-1}^\xi(l_{T-1,0}^o, 0) - V_{T-1}^\xi(l_{T-1,0}^o, 1)] \\
&\partial f_{l_{T-2}^o,1}^\xi / \partial p_{00} = 0, \\
&\partial f_{l_{T-2}^o,0}^\xi / \partial p_{11} = 0, \\
&\partial f_{l_{T-2}^o,1}^\xi / \partial p_{11} = -\gamma[V_{T-1}^\xi(l_{T-1,1}^o, 0) - V_{T-1}^\xi(l_{T-1,1}^o, 1)].
\end{aligned}$$

The feasible set $\mathscr{U}_{T-2}$ does not depend on (p_{00}, p_{11}) and is always nonempty. Assuming the distribution of ξ is nondegenerate (i.e., no component is a constant or a function of other components), the set of values of ξ for which the maximizer exists and is unique has probability 1. Therefore, for almost all ξ, by Danskin's theorem (see [28] Theorem 7.21 or [1] Sect. 4.3.1), the maximum is differentiable at $(p_{00}, p_{11}) \in (0, 1)^2$ with values $\partial V_{T-2}^\xi / \partial p_{ii} = \partial f_{l_{T-2}^o, b_{T-2}}^\xi / \partial p_{ii}$, keeping all decisions fixed to their optimal value. This means that $\partial u_t^e / \partial p_{00} = 0$, $\partial l_t^o / \partial p_{00} = 0$, etc., for all $t \geq T-2$. At times $t \leq T-3$, we have

$$
\begin{aligned}
\partial f^{\xi}_{l^o_t,0}/\partial p_{00} &= \gamma\Big[V^{\xi}_{t+1}(l^o_{t+1,0},0) - V^{\xi}_{t+1}(l^o_{t+1,0},1) \\
&\quad + p_{00}\frac{\partial V^{\xi}_{t+1}(l^o_{t+1,0},0)}{\partial p_{00}} + (1-p_{00})\frac{\partial V^{\xi}_{t+1}(l^o_{t+1,0},1)}{\partial p_{00}}\Big], \\
\partial f^{\xi}_{l^o_t,1}/\partial p_{00} &= \gamma\Big[(1-p_{11})\frac{\partial V^{\xi}_{t+1}(l^o_{t+1,1},0)}{\partial p_{00}} + p_{11}\frac{\partial V^{\xi}_{t+1}(l^o_{t+1,1},1)}{\partial p_{00}}\Big], \\
\partial f^{\xi}_{l^o_t,0}/\partial p_{11} &= \gamma\Big[p_{00}\frac{\partial V^{\xi}_{t+1}(l^o_{t+1,0},0)}{\partial p_{11}} + (1-p_{00})\frac{\partial V^{\xi}_{t+1}(l^o_{t+1,0},1)}{\partial p_{11}}\Big], \\
\partial f^{\xi}_{l^o_t,1}/\partial p_{11} &= \gamma\Big[-V^{\xi}_{t+1}(l^o_{t+1,1},0) + V^{\xi}_{t+1}(l^o_{t+1,1},1) \\
&\quad + (1-p_{11})\frac{\partial V^{\xi}_{t+1}(l^o_{t+1,1},0)}{\partial p_{11}} + p_{11}\frac{\partial V^{\xi}_{t+1}(l^o_{t+1,1},1)}{\partial p_{11}}\Big],
\end{aligned} \tag{41}
$$

and by a similar reasoning one argues that $\partial V^{\xi}_t(l^o_t, b_t)/\partial p_{ii} = \partial f^{\xi}_{l^o_t,b_t}/\partial p_{ii}$, keeping all decisions fixed to their optimal value. One proceeds recursively over t until $\partial V^{\xi}_0(l^o_0, b_0)/\partial p_{ii}$ is reached. Equation (39) expresses (41) in a more compact form.

Finally, we check the conditions for the interchange between expectation and differentiation with respect to (p_{00}, p_{11}) around $(\bar{p}_{00}, \bar{p}_{11}) \in (0,1)^2$, see [28] Theorem 7.44 Assumptions A1, A2, A4:

[A1] $(\bar{p}_{00}, \bar{p}_{11})$ is such that V^{ξ}_0 is well defined and finite. This holds true since

$$
V^{\xi}_0(l_0, b_0) \le \sum_{t=0}^{T-1} \gamma^t C\mathbb{E}\{p^e_t\} + \gamma^T K\mathbb{E}\{p^o_T\} < \infty,
$$
$$
V^{\xi}_0(l_0, b_0) \ge -\sum_{t=0}^{T-1} \gamma^t K\mathbb{E}\{p^o_t\} > -\infty.
$$

[A2] There exists a positive C^{ξ} such that $\mathbb{E}\{C^{\xi}\} < \infty$, and for all (p'_{00}, p'_{11}) in a neighborhood of $(\bar{p}_{00}, \bar{p}_{11})$, and for almost every ξ,

$$
|V^{\xi}_0(l_0, b_0; p'_{00}, p'_{11}) - V^{\xi}_0(l_0, b_0; \bar{p}_{00}, \bar{p}_{11})| \le C^{\xi}\,\|(p'_{00} - \bar{p}_{00}, p'_{11} - \bar{p}_{11})\|.
$$

This holds true since the derivatives in (41) are bounded uniformly over the decisions, using

$$
|V^{\xi}_{t+1}(\cdot,1) - V^{\xi}_{t+1}(\cdot,0)| \le \sum_{k=t+1}^{T-1} \gamma^{k-(t+1)}[G_c p^g_t - O_c p^o_t]^+ + K p^o_t
$$

and using (41) recursively:

$$
\left|\frac{\partial V^{\xi}_t(\cdot,1)}{\partial p_{ii}} - \frac{\partial V^{\xi}_t(\cdot,0)}{\partial p_{ii}}\right| \le \gamma|V^{\xi}_{t+1}(\cdot,1) - V^{\xi}_{t+1}(\cdot,0)| + \gamma\left|\frac{\partial V^{\xi}_{t+1}(\cdot,1)}{\partial p_{ii}} - \frac{\partial V^{\xi}_{t+1}(\cdot,0)}{\partial p_{ii}}\right|.
$$

The finiteness of $\mathbb{E}\{C^{\xi}\}$ results from $\mathbb{E}\{p^o_t\}$ and $\mathbb{E}\{[Cp^e_t - G_c p^g_t]^+\}$ being finite.

[A4] For almost all ξ the function $V^{\xi}(l_t, b_t;\ p_{00}, p_{11})$ is differentiable at $(\bar{p}_{00}, \bar{p}_{11})$. This holds true from $\partial V_t^{\xi}(l_t^o, b_t;\ p_{00}, p_{11})/\partial p_{ii} = \partial f^{\xi}_{l_t^o, b_t}/\partial p_{ii}$ and (41). □

We conclude this section with an observation that leads to a method to select M.

Proposition 5 *In the case* $K = 0$*, the upper bound* $V^{UB}(s_0)$ *in* (35) *is equal to the exact optimal value of the problem. In particular, it is equal to the lower bound* $V^{LB}(s_0)$.

Proof. In the case $K = 0$ (no oil storage capacity), the only decision is whether $u_t^g = 0$ or $u_t^g = 1$ when $b_t = 1$. This decision does not influence the distribution of future states and is only based on the sign of the spark spread $(Cp_t^e - G_c p_t^g)$ at time t. Knowing the price of gas and electricity in advance will in general modify the conditional distributions for the oil price process, but since oil can never be used in the absence of oil storage capacity, this improved information has no impact on the optimal decision policy. Consequently, there is no gain in optimality from perfect foresight assumed within V^{ξ}. □

Proposition 5 is interesting because it provides an accurate method to relate the statistical upper bound to the exact upper bound. In numerical studies, one can select M in (38) by first verifying that the statistical upper bound matches the exact lower bound in the case $K = 0$. When the match is exact to a desired tolerance level, one could also freeze the sample set, given the successful test on $K = 0$.

6 Numerical Test

We evaluate the lower and upper bound and their sensitivities to p_{00}, p_{11} for a particular instance. We set γ to 0.95. We assume that the power generator is operated as a peaker and runs either 0 or 1 hour per day. In our base case, the problem is over $T = 30$ days, with the understanding that the period duration is $\Delta = 1$ hour, corresponding to the hour where the peaker may operate. The decision maker is risk neutral, and thus willing to be fully exposed to spot price variations and physical disruptions. The price of electricity is in \$/MWh, the price of gas is in \$/MMBtu, and the price of oil is in \$/barrel. In the oil price, we neglect the transportation costs and the emission costs. These location-specific components could be added to the oil price in a more detailed study. We assume that one barrel of oil produces 5.5 MMBtu of heat, and that the combustion turbine can convert 1MMBtu of heat to 0.1 MWh of electricity. We neglect the effect of the ambient temperature on efficiency. The power capacity is $C = 100$ MW. The oil storage capacity is set to 3 days of 1-h production by oil, thus $K = 3 \cdot 100 \cdot (1/0.1)/5.5 = 545.5$ barrels. With these values, we also have $G_c = 100 \cdot (1/0.1) = 1000$ and $O_c = 100 \cdot (1/0.1)/5.5 = 181.8$. The price process parameters are as follows:

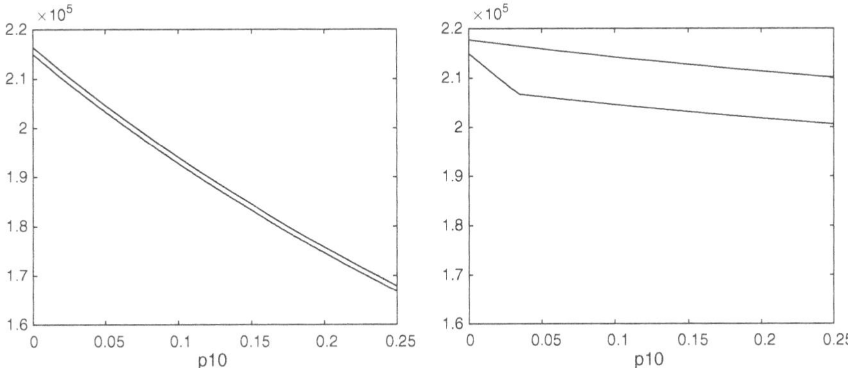

Fig. 1 Statistical upper bound $\hat{V}^{\text{UB}}$ and lower bound V^{LB} on the test problem, as a function of the gas network failure probability $p_{10} = 1 - p_{11}$. Left: $K = 0$ (gas only). Right: $K = 3 \cdot O_c$ (dual-firing)

$$\begin{array}{lll} \log(\zeta^e) = \log(100), & \kappa^e = 0.5, & \sigma^e = 1, \\ \log(\zeta^g) = \log(5), & \kappa^g = 0.3, & \sigma^g = 0.3, \\ \log(\zeta^o) = \log(50), & \kappa^o = 0.1, & \sigma^o = 0.05, \end{array} \quad \Sigma = \begin{bmatrix} 1 & .2 & 0 \\ .2 & 1 & .2 \\ 0 & .2 & 1 \end{bmatrix},$$

where as seen from Σ we have assumed a positive correlation between gas and electric, a positive correlation between gas and oil, and no correlation between oil and electric. Usually oil moves slower than gas, and electric is the most volatile; this is reflected in the volatilities σ.

Following Proposition 5, we do a test run with $K = 0$ (no oil storage) to determine an appropriate value for M, the number of scenarios for the statistical upper bound. $M = 20\,000$ provides a reasonable match. The values of the lower and upper bounds appear in Fig. 1 (Left) and are still distinguishable. The bounds are calculated for various values of the transition matrix P, namely, p_{10} varying from 0 to 0.25, and p_{01} set to 0.85.

Next, we compute the value of the lower bound and the upper bound when $K = 3 \cdot O_c = 545.5$. We report the values as a function of p_{10} as well. The results are depicted in Fig. 1 (Right). On this example, the maximal gap between the two bounds is below 5% of the lower bound value. Finally, we check that the derivatives given in Propositions 3 and 4 are those that are observed on the curves of the figure. (This is indeed the case; those tests are not depicted.)

The comparison between the two cases ($K = 0$, $K = 3 \cdot O_c$) shows the extent to which dual-firing capabilities mitigate the loss of value from the unreliable gas network, as measured here by the transition probability p_{10}.

7 Concluding Remarks

In this paper, we formulate an optimal power production management problem for a dual-firing power generator. We discuss the impact of the gas network reliability on power production. We establish lower bounds and upper bounds that can be used to estimate the benefits of improving the reliability of the gas supply and the benefits of fuel flexibility. In our test the gap is relatively small, indicating that for the price model we considered, the lower bound is based on reasonable assumptions on optimal operations. Otherwise, the lower bound could be improved, for instance by calculating an approximate expected value function at the next state $\widetilde{\mathscr{V}}$ and then estimating on an independent test sample the value of the policy that uses the approximate value function. The upper bound could be improved, for instance, by reintroducing uncertainty on some of the price components and conditioning the distributions on the perfect foresight information.

The stochastic access to the gas network can be interpreted as a random process that enables certain actions at times that are not controlled by the decision maker. The present work relates to the optimal control of such systems. Insights from this work could thus be found relevant to other domains with similar characteristics and reliability concerns.

Acknowledgements This material is based upon work supported by the National Science Foundation under Grant No. 1610825.

References

1. Bonnans, J., Shapiro, A.: Perturbation Analysis of Optimization Problems. Springer, New York (2000)
2. Borovkova, S., Schmeck, M.: Electricity price modeling with stochastic time change. Energy Econ. **63**, 51–65 (2017)
3. Carmona, R., Durrleman, V.: Pricing and hedging spread options. SIAM Rev. **45**(4), 627–685 (2003)
4. Carmona, R., Durrleman, V.: Generalizing the Black-Scholes formula to multivariate contingent claims. J. Comput. Financ. **9**, 42–63 (2005)
5. Deng, S.: Pricing electricity derivatives under alternative stochastic spot price models. In: 33rd Annual Hawaii International Conference on System Sciences (HICSS), pp. 10–20 (2000)
6. Total Electric Power Industry Summary: U.S. Energy Information Administration. https://www.eia.gov/electricity/annual/
7. Escribano, A., Ignacio Peña, J., Villaplana, P.: Modelling electricity prices: international evidence. Oxford Bull. Econ. Stat. **73**(5), 622–650 (2011)
8. Fiacco, A.V., Kyparisis, J.: Convexity and concavity properties of the optimal value function in parametric nonlinear programming. J. Optim. Theory Appl. **48**(1), 95–126 (1986)
9. Fleten, S.E., Näsäkkälä, E.: Gas-fired power plants: investment timing, operating flexibility and CO_2 capture. Energy Econ. **32**(4), 805–816 (2010)
10. Geman, H.: Mean reversion versus random walk in oil and natural gas prices. In: Advances in Mathematical Finance, pp. 219–228. Springer (2007)

11. Golub, G., Meyer Jr., C.: Using the QR factorization and group inversion to compute, differentiate, and estimate the sensitivity of stationary probabilities for Markov chains. SIAM J. Algebr. Discret. Methods **7**(2), 273–281 (1986)
12. Haneveld, W., Stougie, L., van der Vlerk, M.: Simple integer recourse models: convexity and convex approximations. Math. Program. **108**(2), 435–473 (2006)
13. Jaillet, P., Ronn, E., Tompaidis, S.: Valuation of commodity-based swing options. Manag. Sci. **50**(7), 909–921 (2004)
14. Knittel, C., Roberts, M.: Financial models of deregulated electricity prices: an application to the California market. Energy Econ. **27**(5), 791–817 (2005)
15. Koeppel, G., Andersson, G.: The influence of combined power, gas, and thermal networks on the reliability of supply. In: 6th World Energy System Conference, pp. 10–12. Torino, Italy (2006)
16. Kong, N., Schaefer, A., Ahmed, S.: Totally unimodular stochastic programs. Math. Program. **138**(1–2), 1–13 (2013)
17. Laporte, G., Louveaux, F.: The integer L-shaped method for stochastic integer programs with complete recourse. Oper. Res. Lett. **13**(3), 133–142 (1993)
18. Li, T., Eremia, M., Shahidehpour, M.: Interdependency of natural gas network and power system security. IEEE Trans. Power Syst. **23**(4), 1817–1824 (2008)
19. Lien, D.: Moments of ordered bivariate log-normal distributions. Econ. Lett. **20**(1), 45–47 (1986)
20. Lucia, J., Schwartz, E.: Electricity prices and power derivatives: evidence from the Nordic power exchange. Rev. Deriv. Res. **5**(1), 5–50 (2002)
21. Margrabe, W.: The value of an option to exchange one asset for another. J. Financ. **33**(1), 177–186 (1978)
22. Munoz, J., Jimenez-Redondo, N., Perez-Ruiz, J., Barquin, J.: Natural gas network modeling for power systems reliability studies. In: Proceedings of the 2003 IEEE Power Tech Conference, pp. 1–8. Bologna, Italy (2003)
23. Special Reliability Assessment: Potential bulk power system impacts due to severe disruptions on the natural gas system. North American Electric Reliability Corporation (2017)
24. Ross, S., Zhu, Z.: On the structure of a swing contract's optimal value and optimal strategy. J. Appl. Probab. **45**(1), 1–15 (2008)
25. Schwartz, E.: The stochastic behavior of commodity prices: implications for valuation and hedging. J. Financ. **52**(3), 923–973 (1997)
26. Seneta, E.: Sensitivity of finite Markov chains under perturbation. Stat. Probab. Lett. **17**(2), 163–168 (1993)
27. Shapiro, A.: Analysis of stochastic dual dynamic programming method. Eur. J. Oper. Res. **209**(1), 63–72 (2011)
28. Shapiro, A., Dentcheva, D., Ruszczyński, A.: Lectures on stochastic programming: modeling and theory. SIAM, Philadelphia, PA (2009)
29. Sun, R., Shylo, O., Schaefer, A.: Totally unimodular multistage stochastic programs. Oper. Res. Lett. **43**(1), 29–33 (2015)
30. Webster, M., Schmalensee, R.: Growing concerns, possible solutions: the interdependency of natural gas and electricity systems. Technical report, MIT Energy Initiative (2014)

Efficient Piecewise Linearization for a Class of Non-convex Optimization Problems: Comparative Results and Extensions

Giorgio Fasano and János D. Pintér

Abstract This research work originates from a challenging control problem in space engineering that gives rise to hard nonlinear optimization issues. Specifically, we need the piecewise linearization (PL) of a large number of non-convex univariate functions, within a mixed integer linear programming (MILP) framework. For comparative purposes, we recall a well-known classical PL formulation, an alternative approach based on disaggregated convex combination (DCC), and a more recent approach proposed by Vielma and Nemhauser. Our analysis indicates that—in the specific context of our study—the DCC-based approach has computational advantages: this finding is supported by experimental results. We discuss extensions and variations of the basic DCC paradigm. Extensions to a number of possible application areas in robotics and automation are also envisioned.

Keywords Separable functions · Piecewise linearization of non-convex univariate and multivariate functions · Large-scale mixed integer linear programming · Nonlinear programming · Global optimization · Comparative numerical experiments · Control problems · Control dispatch optimization · Space engineering and other applications

MSC Classification (2010) 90C06 · 90C11 · 90C26 · 90C30 · 90C59 · 90C90

1 Introduction

The topic presented here originates from a research activity related to space engineering, in the context of handling a challenging optimal control dispatch problem. The underlying class of problems is related to the distributed control of a dynamic

G. Fasano (✉)
Thales Alenia Space, Turin, Italy
e-mail: giorgio.fasano@thalesaleniaspace.com

J. D. Pintér
Lehigh University, Bethlehem, PA, USA
e-mail: jdp416@lehigh.edu

J. D. Pintér and T. Terlaky (eds.), *Modeling and Optimization: Theory and Applications*, Springer Proceedings in Mathematics & Statistics 279,
https://doi.org/10.1007/978-3-030-12119-8_3

system through a set of actuators. This problem-class can be handled by a new modeling approach [4, 5], using MILP tools. Different model versions can be considered, depending on the objective function form chosen. Our present discussion focuses on the case when these functions are separable and consist of non-convex univariate terms. Piecewise linearization is used to obtain mixed integer linear programming formulations that approximate these highly nonlinear models.

A well-known approach to PL is based on Special Ordered Sets of type 2 (SOS2) proposed by Beale and Tomlin [2], Beale and Forrest [3]. Let us recall that SOS2 is a set of continuous or integer variables within which at most two variables can have nonzero values; it is also required that these nonzero variables are adjacent according to the ordering defined on the set. The SOS2 concept allows the treatment of piecewise linearization algorithmically, within a tailored branch and bound procedure: cf. also Williams [17].

Currently, several efficient model-based PL techniques are available. These methods are based on introducing additional variables and constraints: consult, e.g., the reviews by Vielma et al. [14], and by Lin et al. [7]. The approaches proposed by Vielma et al. [14], and Vielma and Nemhauser [15]—differently from previous PL strategies—require only a logarithmic number of additional binary variables and constraints, for both univariate and multivariate functions. Although further experimental analysis should be carried out in order to compare the various PL methods, it is deemed to be the best choice, especially when the linearized functions have a large number of breakpoints (polytopes in the multivariate case). At the same time, if the number of breakpoints is rather limited, then non-logarithmic approaches are preferred: consult Vielma et al. [14].

On the basis of a comparative analysis discussed in our present study, we propose the so-called *disaggregated convex combination* (DCC) method (see, e.g., [14]), as a well-suited approach to the class of problems studied here. According to our analysis, the DCC method deals efficiently with a large number of functions, assuming a limited number of breakpoints. If the number of breakpoints per function increases, then a drastic reduction of the computational performance is expected. In order to enhance the practical applicability of the DCC paradigm, an ad hoc variation is also proposed. This is based on the consideration that a function with many breakpoints can be partitioned into a set of functions with only a few of these. A new version of the DCC method for multivariate functions, involving a reduced number of additional variables, is also outlined as a promising extension towards a range of applications.

Following this Introduction, in Sect. 2, we briefly review the classical PL approach based on convex combinations. We will refer to this as approach as CL, cf., e.g., Taha [12] or Williams [17]. This is followed by reviewing the DCC method and the technique proposed by Vielma and Nemhauser [15], referred to as VN. Section 3 describes the class of problems the present study focuses on, pointing out its special structure and the suitability of the DCC formulation. Section 4 presents comparative experimental results, obtained by applying the CL, VN and DCC formulations. Possible DCC extensions are discussed in Sect. 5.

2 Piecewise Linearization Methods: A Brief Review

Consider a continuous nonlinear univariate function $f(u)$ defined on the finite interval $[a, b]$. In order to approximate this function by piecewise linear segments, a monotonically increasing sequence of breakpoints $\{a_k\}$ for $k = 0, 1, \ldots, m$ is introduced, defining $a = a_0$ and $b = a_m$. Next, we evaluate the function at all breakpoints, and connect the adjacent points $(a_{k-1}, f(a_{k-1}))$ and $(a_k, f(a_k))$ for $k = 1, \ldots, m$ by line segments. To illustrate this conceptually simple procedure by an example, see Fig. 1 which shows the function $f(x) = 0.2\,x + sin(x + 3) + 2\,cos(1 + 2\,x)$ defined over the interval $x \in [1, 10]$, together with its piecewise linear approximation based on the input argument values 1, 2, …, 10 and the corresponding function values.

The quality of PL approximations depends on the function f, as well as on the number and location of the breakpoints. For our discussion, we introduce the general family of PL functions Λ defined for a given sequence of breakpoints $\{a_k\}$ and multipliers $\{\lambda_k\}$ as

$$\Lambda\big[f(u)\big] = \sum_{k=0}^{m} f(a_k)\lambda_k, \tag{1}$$

$$u = \sum_{k=0}^{m} a_k\lambda_k, \tag{2}$$

$$\sum_{k=0}^{m} \lambda_k = 1. \tag{3}$$

Here, $\lambda_k \geq 0$ for $k = 0, 1, \ldots, m$, and by assumption only two adjacent λ_k, λ_{k+1} values can be positive. Following the model formulation discussed, e.g., by Williams

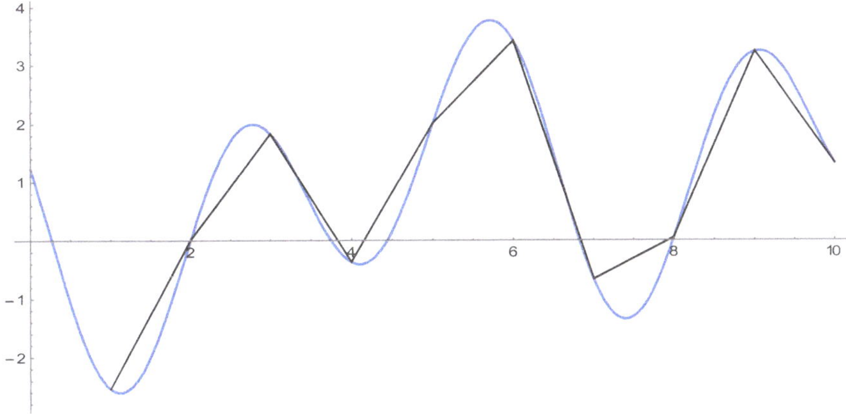

Fig. 1 Piecewise linear approximation of a nonlinear function

[17], this adjacency condition can be expressed by introducing $m + 1$ binary variables $\delta_k \in \{0, 1\}$ and the additional constraints shown below:

$$\lambda_0 \leq \delta_0, \tag{4.1}$$

$$\lambda_k \leq \delta_{k-1} + \delta_k \ \forall k \in \{1, \ldots, m-1\}\ , \tag{4.2}$$

$$\lambda_m \leq \delta_{m-1}, \tag{4.3}$$

$$\sum_{k=0}^{m-1} \delta_k = 1. \tag{4.4}$$

This modeling approach requires $m + 1$ binary variables, $m + 1$ continuous variables and $m + 5$ constraints, as shown by (1) to (4.4). This formulation will be denoted in the following by CL.

As already mentioned, alternative model formulations [7, 14, 15] have also been studied to deal with the adjacency condition.

Next, the DCC model is presented, with reference to the univariate function type $f(u)$ introduced above. Specifically, Eqs. (1), (2), and (3) are replaced by (5), (6), and (7) and the added constraints (8):

$$\Lambda\left[f(u)\right] = \sum_{k=0}^{m-1} (f(a_k)\lambda_k + f(a_{k+1})\widehat{\lambda}_k), \tag{5}$$

$$u = \sum_{k=0}^{m-1} (a_k\lambda_k + a_{k+1}\widehat{\lambda}_k), \tag{6}$$

$$\sum_{k=0}^{m-1} (\lambda_k + \widehat{\lambda}_k) = 1, \tag{7}$$

$$\lambda_k + \widehat{\lambda}_k = \sigma_k \ \text{ for } k \in \{0, 1, \ldots, m-1\}, \ \text{ where } \lambda_k, \widehat{\lambda}_k \geq 0 \text{ and } \sigma_k \in \{0, 1\}. \tag{8}$$

Observe that, compared to the CL method, the number of additional continuous variables has nearly doubled. More precisely, relations (5)–(8) require altogether $2\,m$ continuous variables, m binary variables, and $m + 3$ constraints.

Next, the VN approach is described. In addition to (1)–(3), the inequalities shown below are introduced:

$$\sum_{p \in S^{+}(k)} \lambda_p \leq \rho_k, \tag{9.1}$$

$$\sum_{p \in S^{-}(k)} \lambda_p \leq 1 - \rho_k. \tag{9.2}$$

Here, $\lambda_p \geq 0$, $p \in P = \{0, 1, \ldots, m\}$, $\rho_k \in \{0, 1\}$, $k \in \{1, \ldots, \lceil \log_2 m \rceil\}$; and the sets $S^+(k)$, $S^-(k)$ are defined as follows. First, an injective function $B : \{1, 2, \ldots, m\} \to \{0, 1\}^{\lceil \log_2 m \rceil}$ is considered. Here, by assumption, the vectors $B(p)$ and $B(p+1)$ differ in at most one of their components, for all values $p \in \{1, 2, \ldots, m-1\}$. Next, vector $B(p) = (\rho_1, \ldots, \rho_{\lceil \log_2 m \rceil})$ is chosen with the added condition $B(0) = B(1)$. Then, the sets $S^+(k)$ and $S^-(k)$ are defined by the relations:

$$S^+(k) = \{p | p \in \{1, \ldots, m-1\}, B(p)_k = B(p+1)_{k+1} = 1\} \cup \{p | p \in \{0, m\}, B(p)_k = 1\}, \quad (9.3)$$

$$S^-(k) = \{p | p \in \{1, \ldots, m-1\}, B(p)_k = B(p+1)_k = 0\} \cup \{p | p \in \{0, m\}, B(p)_k = 0\}, \quad (9.4)$$

where $B(p)_k$ is the kth component of vector $B(p)$.

To illustrate the VN approach, we give an example. An interval with 11 breakpoints is considered and the following injective function $B : \{1, 2, \ldots, 10\} \to \{0, 1\}^4$, satisfying the stated assumptions, is introduced:

$$\begin{aligned}
&B(0) = B(1) = (0, 0, 0, 0), \quad B(2) = (0, 0, 0, 1), \quad B(3) = (0, 0, 1, 1),\\
&B(4) = (0, 0, 1, 0), \quad B(5) = (0, 1, 1, 0), \quad B(6) = (0, 1, 1, 1), \quad B(7) = (0, 1, 0, 1),\\
&B(8) = (0, 1, 0, 0), \quad B(9) = (1, 1, 0, 0), \quad B(10) = (1, 1, 0, 1).
\end{aligned}$$

This gives rise to the sets $S^+(k)$ and $S^-(k)$ shown below:

$$\begin{aligned}
&S^+(1) = \{9, 10\}, S^+(2) = \{5, 6, 7, 8, 9, 10\}, S^+(3) = \{3, 4, 5\},\\
&S^+(4) = \{2, 6, 10\},\\
&S^-(1) = \{0, 1, 2, 3, 4, 5, 6, 7\}, S^-(2) = \{0, 1, 23,\}, S^-(3) = \{0, 1, 7, 8, 9, 10\},\\
&S^-(4) = \{0, 4, 8\}
\end{aligned}$$

Now, conditions (9.1) and (9.2) read as follows:

$$\begin{aligned}
&\lambda_9 + \lambda_{10} \leq \rho_1,\\
&\lambda_5 + \lambda_6 + \lambda_7 + \lambda_8 + \lambda_9 + \lambda_{10} \leq \rho_2,\\
&\lambda_3 + \lambda_4 + \lambda_5 \leq \rho_3,\\
&\lambda_2 + \lambda_6 + \lambda_{10} \leq \rho_4,\\
&\lambda_0 + \lambda_1 + \lambda_2 + \lambda_3 + \lambda_4 + \lambda_5 + \lambda_6 + \lambda_7 \leq 1 - \rho_1,\\
&\lambda_0 + \lambda_1 + \lambda_2 + \lambda_3 \leq 1 - \rho_2,\\
&\lambda_0 + \lambda_1 + \lambda_7 + \lambda_8 + \lambda_9 + \lambda_{10} \leq 1 - \rho_3,\\
&\lambda_0 + \lambda_4 + \lambda_8 \leq 1 - \rho_4.
\end{aligned}$$

The ingenious approach proposed by Vielma and Nemhauser [15] is, in general, very efficient. The VN method reduces the number of additional binary variables and constraints significantly. Indeed, to achieve a piecewise linearization with m line segments, this modeling approach requires only $\lceil \log_2 m \rceil$ binary variables, $m + 1$ continuous variables and $3 + 2\lceil \log_2 m \rceil$ constraints. To illustrate this point, assume that $m = 1000$: then the CL model development approach requires 1001 binary and 1001 continuous variables, plus 1005 constraints; while the VN method requires only 10 binary and 1001 continuous variables, plus 25 constraints. As it is well known, both the theoretical and numerical difficulty of MILP models is determined primarily by the number of binary variables and binary (or mixed integer) constraints, while the number of continuous variables has only a secondary influence.

3 An Illustrative Case Study

As mentioned, this research originates from a space engineering study, concerning the attitude control of spacecraft [1, 4, 5]. In this context, a dedicated controller has the scope of determining the overall control action, aimed at determining step-by-step the expected system attitude. A number of thrusters are available to exert the overall force and torque required. In demanding missions, the control designer has the difficult task of appropriately determining the layout of these thrusters, since different configurations can lead to significantly different overall performance, in terms of propellant consumption. The thruster orientation—assumed to be kept fixed during the entire mission—is, hence a major issue. This control problem, expressed in a general form, represents the case study discussed in this section. Therefore, a dynamic system (assumed to be a rigid body) is considered, and its overall control is performed by a number of actuators.

The NLP model based on time discretization [5] is presented next. For this purpose, the sets A and I, relevant, respectively, to the actuators and instants (precisely given moments of time) are introduced. Given the system-based reference frame $O(x, y, z)$, the following notations are adopted:

u_{ri} is, for each actuator r, the Euclidean norm of the force exerted at instant i, $\boldsymbol{v}_r = (v_{rx}, v_{ry}, v_{rz})$ is the unit vector (i.e., with norm 1), representing the orientation of each actuator r, $\boldsymbol{F}_i = (F_{rxi}, F_{ryi}, F_{rzi})$ and $\boldsymbol{T}_{\mathrm{i}} = (T_{xi}, T_{yi}, T_{zi})$ are the force and torque, respectively, representing the overall control required, $\boldsymbol{P}_r = (P_{rx}, P_{ry}, P_{rz})$ is, for each actuator r, the application-point vector (constant) of the force functions $u_{ri}\boldsymbol{v}_r$ that holds for all instants i, $\underline{U}_r$, $\overline{U}_r$, $\underline{V}_r$,$\overline{V}_r$ and L_r are the related technological bounds.

Then, we can consider the following NLP model:

$$min \sum_{\substack{r \in A \\ i \in I}} f_r(u_{ri}), \tag{10}$$

subject to

$$\forall i \in I \quad \begin{pmatrix} v \\ P \times v \end{pmatrix} \begin{pmatrix} u_{1i} \\ \cdots \\ u_{ri} \\ \cdots \\ u_{N_A i} \end{pmatrix} = \begin{pmatrix} F_i \\ T_i \end{pmatrix}, \tag{11}$$

$$\forall r \in A \quad v_{rx}^2 + v_{ry}^2 + v_{rz}^2 = 1, \tag{12}$$

$$\forall r \in A \ \forall i \in \{0, 1, \ldots, |I|-2\} \quad |u_{r(i+1)} - u_{ri}| \leq L_r, \tag{13}$$

where $\forall r \in A \quad u_{ri} \in \left[\underline{U}_r, \overline{U}_r\right], \underline{V}_r \leq (v_{rx}, v_{ry}, v_{rz})^T \leq \overline{V}_r$.

The univariate functions $f_r(u_{ri})$ in (10) are non-convex and the system of equations (11) expresses the assigned system control. In a more explicit (vector) formulation, the latter can be written as

$$\forall i \in I \ \sum_{r \in A} u_{ri} \boldsymbol{v}_r = \boldsymbol{F}_i, \tag{14.1}$$

$$\forall i \in I \ \sum_{r \in A} \boldsymbol{P}_r \times (u_{ri} \boldsymbol{v}_r) = \boldsymbol{T}_i. \tag{14.2}$$

Equations (12) are normalization constraints. Equations (13) are "uniformity" conditions on the variables u_{ri} (stating that the maximum allowable difference of the forces applied by the same actuator in two subsequent instants cannot exceed a given threshold).

The optimization model expressed by (10)–(14.1, 14.2) has a non-convex objective function, subject to a set of bilinear and quadratic equations, in addition to "easy-to-handle" linear inequalities. Therefore, it is very hard to solve, especially for large-scale instances. This has led to proposing an ad hoc heuristic solution methodology, with a global optimization (GO) flavor (Fasano 2018). Namely, two subproblems, substituting the original one, are solved recursively until a satisfactory solution is found. In the first subproblem, focusing mainly on the orientation of the actuators, we solve problem (10)–(14.1, 14.2) only with respect to a limited subset of time moments, assumed to be representative of the whole time period considered. This problem can be tackled directly by means of a GO solver such as ANTIGONE [8], BARON [13], LaGO [9] or LGO [10, 11]; or, alternatively, it can be approximated as an MILP problem through a suitable discretization of variables v. The second subproblem consists of optimizing the overall problem (10)–(14.1, 14.2), covering the entire set of instants I, once the variables v have been assigned as solutions of the first subproblem.

Since the solution of the first subproblem involves exclusively small-scale instances, in this work we shall consider only the second subproblem, denoted by P. In order to illustrate its general structure, it is formulated in a compact form. For this purpose, the sets of indices H, I, and J are introduced: postulating that the cardinality of J is smaller than that of H, i.e., $|J| < |H|$. Then problem P is stated as follows:

$$min \sum_{\substack{h \in H \\ i \in I}} f_{hi}(u_{hi}) \tag{15}$$

Subject to

$$\forall i \in I \quad Q_i(u_{hi}) = C_i, \tag{16}$$

$$\forall h \in H \quad \forall i \in I : i \leq |I|-2 \quad |u_{hi} - u_{h,i+1}| \leq L_h. \tag{17}$$

Here, $\forall i \in I \quad Q_i \in R^{|J| \times |H|}$ (matrices of constants), $\forall i \in I \quad C_{i1}, \ldots, C_{i|J|} \in R^{|J|}$ (vectors of constants), $\forall h \in H \quad \forall i \in I \quad \underline{U}_h \leq u_{hi} \leq \overline{U}_h$, with both $\underline{U}_h$ and $\overline{U}_h \geq 0$; and we postulate that $L_h > 0$. The objective function defined by (5) is separable, consisting of the sum of $|H||I|$ non-convex univariate functions $f_{hi}(u_{hi})$. For each $h \in H$, it is supposed that all the corresponding functions $f_{hi}(u_{hi})$ are identical to each other, independently from the considered instant $i \in I$: this is assumed solely for reasons of conformity with the original real-world problem. More precisely, this statement is expressed by

$$\forall h \in H \quad \forall i, j \in I \quad \forall u \in [\underline{U}_h, \overline{U}_h] \qquad f_{hi}(u) = f_{hj}(u). \tag{18}$$

Before presenting our experimental analysis, let us note that real-world problem instances relevant to the studied application typically have the following features: a large number of nonlinear functions f_{hi} are required (their number often exceeds 1000); each function f_{hi} can be adequately approximated with a fairly limited number of breakpoints (usually, about 10 breakpoints are sufficient). These two facts indicate that the large-scale nature of the problem is mainly due to the number of nonlinear functions present, rather than to the required PL aspect. Therefore—rather differently from numerous other real-world applications—the number of breakpoints for each linearized function (and, consequently, the number of additional variables and constraints) does not represent a major hindrance per se.

Dealing with this specific problem-type, the CL method presented in Sect. 2 is rather inefficient. We also observed that, for certain real-world instances of our control engineering problem, the VN formulation does not lead to satisfactory performance. Both of these observations are supported by our detailed test presented in Sect. 4. Therefore, the DCC method has been considered as a promising PL alternative for the class of problems in question.

Let us point out first that (differently from the CL and VN approaches) here all constraints are equations: in a MILP model, these are often easier to deal with than the corresponding inequalities. This formulation (as well as CL and other methods utilizing m binary variables) is suitable to consider relations between the independent variables u_{hi} associated with the functions $f_{hi}(u_{hi})$. We illustrate this point by the following example:

$$\sigma_{hik} \leq \sigma_{h,i-1,k-1} + \sigma_{h,i-1,k} + \sigma_{h,i-1,k+1}, \sigma_{hik} \leq \sigma_{h,i+1,k-1} + \sigma_{h,i+1,k} + \sigma_{h,i+1,k+1}. \quad (19)$$

Such constraints represent discretized "neighborhood" conditions, similar to the continuous ones expressed by (16). Note also that this approach can be applied easily, since—as opposed to other approaches including SOS2—each binary variable σ_{hik} acts directly upon the corresponding discretization subinterval. Finally, we note that similar to previously studied PL methods, cf., e.g., Li and Yu [6], the binary variables need to be introduced exclusively in association with the non-convex portions of $f(u)$. This can be done easily: in the DCC formulation, it is sufficient to ignore conditions (8) for all subintervals $[a_r,a_s]$ in $[a_0,a_m]$ where $f(u)$ is convex, and set $\widehat{\lambda}_k = 0$ for all the corresponding indices k.

4 Comparative Experimental Analysis

In order to compare the performance of the three PL methods discussed here, we conducted a set of numerical tests, based on a number of instances of problem P. The first test set suite consists of 90 instances, referred to as test set 1–90. It has been built as follows: based on the underlying control engineering application, we considered 15 different types of nonlinear functions $f_1(u), \ldots, f_{15}(u)$ with 11 breakpoints each (thereby $m = 10$), defined over the interval $[\underline{U}, \overline{U}]$. To illustrate, Fig. 2 shows three appropriately scaled instances of the test functions used in our comparative numerical study.

For simplicity, it has been assumed that $\forall h, l \in H, \forall i \in I, \forall u \in [\underline{U}, \overline{U}] \quad f_{hi}(u) = f_{li}(u)$. We used two different sets of vectors C_{i1} and C_{i2} and three different cardinalities for the set I, defining $|I_1| = 50$, $|I_2| = 75$ and $|I_3| = 250$. Table 1 summarizes the basic structure of our test model instances.

In the next test suite, referred to as testset 91–180, 90 new problem instances have been generated, setting $|I| = 100$. For this purpose, eight new sets of vectors C_i have been considered. These have been selected, in an attempt to generate worst case

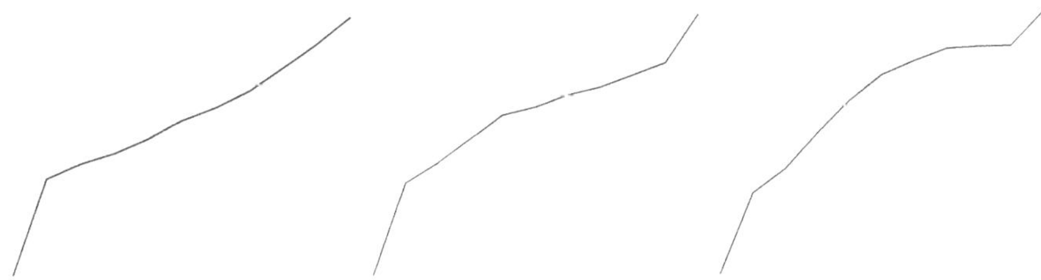

Fig. 2 Examples of test functions used in our study

Table 1 Test set 1–90: summary of overall structure

Test subsets	$f(u)$	C_i	I
1–15	f_1–f_{15}	C_{i1}	I_1
16–30	f_1–f_{15}	C_{i2}	I_1
31–45	f_1–f_{15}	C_{i1}	I_2
46–60	f_1–f_{15}	C_{i2}	I_2
61–75	f_1–f_{15}	C_{i1}	I_3
76–90	f_1–f_{15}	C_{i2}	I_3

instances for the DCC approach. In a third set of experiments, a further comparison of VN and DCC has been made, reconsidering the test subset 1–15 but using $|I|= 350$.

All PL intervals taken into account in the tests discussed have 11 breakpoints. Following Vielma and Nemhauser [15], in the VN method implementation a Gray binary code (cf., e.g., [16]) has been adopted, in order to build an appropriate injective function $B : \{1, 2, \ldots, 10\} \rightarrow \{0, 1\}^4$. Specifically, the following settings have been used:

$$B(1) = (0, 0, 0, 0), B(2) = (0, 0, 0, 1), B(3) = (0, 0, 1, 1), B(4) = (0, 0, 1, 0),$$
$$B(5) = (0, 1, 1, 0), B(6) = (0, 1, 1, 1), B(7) = (0, 1, 0, 1), B(8) = (0, 1, 0, 0),$$
$$B(9) = (1, 1, 0, 0), B(10) = (1, 1, 0, 1).$$

The DCC method has been applied to all tests presented in this section, without taking advantage of the distinction between convex and non-convex segments of the linearized functions (for the sake of simplicity).

All tests reported here have been conducted using IBM ILOG CPLEX 12.3, installed on a personal computer, equipped with a Core 2 Duo P8600, 2.40 GHz processor and 1.93 GB RAM; running under MS Windows XP Professional, Service Pack 2.

A computational study of test set 1–90 has been performed applying all three methods discussed here. The computational time limit of 300 s has been set for each individual test. If—due to the computational time limit imposed—only suboptimal solutions have been found, then the actual computational time necessary to reach the best solution is considered instead of the computational time period of 300 s. If no solution has been found, then the computational time 300 s has been reported. This criterion, together with the same computational time limit, has been used in all tests. Table 2 shows the results obtained reporting the total computational time used, the number of instances not solved within the computational time limit imposed, and the number of proven optimal solutions.

For test suite 1–90, the DCC method was outperformed by the VN method only in one case (in terms of computational time), and it obtained the same objective function value in 12 cases. The CL approach was faster than the DCC method in 14 test instances, however, with less than 10 s runtime difference in each such case.

Considering the entire test suite, the DCC method clearly outperforms both CL and VN, in terms of total computational time, the number of solutions found, and the number of provably optimal solutions found. The optimality of a solution in a MILP instance is detected by the solver when the node tree has been exhaustively explored. It is understood that if no proven optimal solution is explicitly declared, then this means that a number of nodes, potentially up to improving the last solution found, are still pending (and that they have not been explored due to the assigned computational time limit). Similarly, in our tests, when it is declared that no result has been found, it simply means that the computational time limit has been reached before obtaining any integer-feasible solution. All instances considered in this section are a priori known to be integer-feasible.

A predictable difficulty is the instance scale that will have an influence on the solution computational time: in general, the higher the cardinality of I is, the more computational time is required to obtain a solution. It was also observed that changes in the sets of vectors C_{i1} and C_{i2} may also give rise to significant differences in terms of computational performance.

A comparison of the VN and DCC methods for this test set is reported in Table 3, applying the abovementioned timing criterion. Recall that we did not use the CL approach in the following tests, due to its perceived underperformance.

Concerning this set of tests, in 14 cases for which no results were found with the DCC method, the VN method found a solution for each, without proving optimality. In 6 cases VN outperformed DCC, with respect to the computational time required (with computational time differences less than 10 s, except for one test in which the gap reached 142 s, while finding the same value for the objective function). However, considering the entire test suite, DCC again outperforms VN in terms of computational time, number of solutions found, and number of provably optimal solutions. The corresponding instances are referred to as test suite 181–195. Our computational results are reported in Table 4.

Table 2 Test set 1–90: summary results

Method	Total computational time seconds	No. of model instances solved	No. of proven optimal solutions
CL	10,261	70	32
VN	8223	72	34
DCC	1855	90	60

Table 3 Tests 91–180: summary results

Method	Total computational time seconds	No. of model instances solved	No. of proven optimal solutions
VN	20,917	39	10
DCC	17,360	56	28

Table 4 Tests 181–195: summary results

Method	Total computational time seconds	No. of model instances solved	No. of proven optimal solutions
VN	3354	4	4
DCC	1670	11	2

In three cases of set 181–195, VN outperformed DCC, in terms of computational time (with a maximal computational time difference of 143 s), while finding the same objective function value. In one of these cases, VN—differently from DCC—succeeded to prove the optimality of the solution found. However, considering the entire test suite, DCC again outperforms VN both in terms of computational time and the number of solutions found, although not in terms of number of provably optimal solutions.

Quite unexpectedly, a certain complementarity between VN and DCC has been observed for test suites 91–180 and 181–195 and this aspect could be the subject of future research. In most cases, where the first approach was not able to find a solution, the second did; conversely, the aforementioned 14 cases of set 91–180 were solved only by VN. In the tests, 181–195, the 4 unsuccessful tests for DCC were also solved by VN. Table 5 summarizes the matrix dimensions relative to test suite 1–90 for CL.

Table 6 reports the matrix dimensions associated with all three test suites for the VN method.

Finally, Table 7 reports the matrix dimensions associated with all three test suites for the DCC method proposed here.

Table 5 Matrix dimensions: CL method

Test sets	Rows	Columns	Nonzeros	0–1 variables
1–30	6819	9589	30,365	4080
31–60	10,169	14,289	45,265	6080
61–90	33,619	47,189	149,565	20,080

Table 6 Matrix dimensions: VN method

Test sets	Rows	Columns	Nonzeros	0–1 variables
1–30	5187	7141	30,773	1632
31–60	7737	10,641	45,873	2432
61–90	25,587	35,141	15,1573	8032
91–180	10,287	14,141	60,973	3232
181–195	35,787	49,141	211,973	11,232

Table 7 Matrix dimensions: DCC method

Test set	Rows	Columns	Nonzeros	0–1 variables
1–30	6003	13,261	33,221	4080
31–60	8953	19,761	49,521	6080
61–90	29,603	65,261	163,621	20,080
91–180	11,903	26,261	65,821	8080
181–195	41,403	91,261	228,821	28,080

5 Possible Extensions

In our case study, the demonstrated advantage of DCC with respect to VN is deemed to be primarily related to the fact that—despite the significant amount of non-convex functions required—their PL approximations are based on a rather limited number of breakpoints. It seems reasonable to conjecture that the larger the number of breakpoints becomes, the less advantage can be expected. An obvious way to reduce this effect consists of partitioning functions that are approximated through a large number of breakpoints into sets of functions with a more limited number of breakpoints. In the following, an extension of the DCC method to such partitioned functions is presented.

To this end, consider a general piecewise linear continuous univariate function $f(u)$ defined over the interval $[a_0, a_m]$ by the points $(a_k, f(a_k))$, for $k \in S = \{0, 1, \dots, m\}$. A collection of ordered subsets of S (inheriting the same ordering), denoted as S_l, are introduced as follows:

$$S_l = \left\{k \in S | k_{l0} < k \leq k_{lm_l}\right\} = \left\{k_{l0}, k_{l1}, \dots, k_{lm_l} - 1, k_{lm_l}\right\},$$
$$l = 1, \dots, L, k_{l0}, k_{lm_l} \in S,$$

$$\forall l \; |S_l| \geq 2,$$

$$k_{1,0} = 0, k_{Lm_L} = m,$$

for any two subsets S_l and S_{l+1} $(l = 1, \dots, L-1)$ the last element of S_l (i.e.k_{lm_l}) is equal to the first element of S_{l+1} (i.e.,$k_{l+1,0}$).

Then the following formulation, corresponding to (5)–(8) holds:

$$f(u) = \sum_{l=1}^{L} \sum_{k=k_{l0}}^{k_{lm_l} \; 1} (f(a_k)\lambda_k + f(a_{k+1})\widehat{\lambda}_k), \qquad \lambda_k, \widehat{\lambda}_k \geq 0, \tag{20}$$

$$u = \sum_{l=1}^{L} \sum_{k=k_{l0}}^{k_{lm_l}-1} (a_k\lambda_k + a_{k+1}\widehat{\lambda}_k), \tag{21}$$

$$\forall l \quad \sum_{k=k_{l0}}^{k_{lm_l}-1} (\lambda_k + \widehat{\lambda}_k) = \hat{\sigma}_l, \quad \hat{\sigma}_l \in \{0, 1\}, \tag{22}$$

$$\forall l \quad \forall k \in \left\{k_{l0}, .., k_{lm_l} - 1\right\} \lambda_k + \widehat{\lambda}_k = \sigma_{lk}, \quad \sigma_{lk} \in \{0, 1\}, \tag{23}$$

$$\sum_{l=1}^{L} \hat{\sigma}_l = 1. \tag{24}$$

Let us point out that conditions (20)–(24) reduce to (5)–(8), with respect to the subinterval $[a_{k_{l0}}, a_{k_{lm_l}}] \subset [a_0, a_m]$, corresponding to the (unique) index l for which $\hat{\sigma}_l = 1$ holds. When solving a MILP model involving formulation (20)–(24), a branch and bound strategy prioritizing variables $\hat{\sigma}_l$ over variables σ_{lk} can be applied.

Additionally, a variation of the DCC method that allows for a significant reduction of the number of binary variables is proposed for the case of continuous multivariate functions. The corresponding discussion is introduced by considering the bivariate case, cf. Williams [17] that can be of practical interest, for instance, when bilinear problems are involved. For this purpose, the univariate function $f(u)$ is replaced by the bivariate function $f(u_1, u_2)$. Assuming that $u_1 \in [a_0, a_m]$ and $u_2 \in [b_0, b_p]$, we can introduce the increasing sequence of breakpoints a_k for $k = 0, 1, \ldots, m$ and b_h for $h = 0, 1, \ldots,$ p. A direct application of the DCC methods is as follows:

$$\Lambda\left[f(u_1, u_2)\right] = \sum_{k=0}^{m-1} \sum_{h=0}^{p-1} (f(a_k, b_h)\lambda_{kh} + f(a_{k+1}, b_h)\widehat{\lambda}_{kh} + f(a_{k+1}, b_{h+1})\mu_{kh} + f(a_k, b_{h+1})\widehat{\mu}_{kh}), \tag{25}$$

$$u_1 = \sum_{k=0}^{m-1} \sum_{h=0}^{p-1} (a_k\lambda_{kh} + a_{k+1}\widehat{\lambda}_{kh} + a_{k+1}\mu_{kh} + a_k\widehat{\mu}_{kh}), \tag{26}$$

$$u_2 = \sum_{k=0}^{m-1} \sum_{h=0}^{p-1} (b_k\lambda_{kh} + b_k\widehat{\lambda}_{kh} + b_{k+1}\mu_{kh} + b_{k+1}\widehat{\mu}_{kh}), \tag{27}$$

$$\sum_{k=0}^{m-1} \sum_{h=0}^{p-1} (\lambda_{kh} + \widehat{\lambda}_{kh} + \mu_{kh} + \widehat{\mu}_{kh}) = 1, \tag{28}$$

$$\forall k \in \{0, \ldots, m-1\} \quad \forall h \in \{0, \ldots, p-1\} \quad \lambda_{kh} + \widehat{\lambda}_{kh} + \mu_{kh} + \widehat{\mu}_{kh} = \sigma_{kh}, \lambda_{kh}, \widehat{\lambda}_{kh}, \mu_{kh}, \widehat{\mu}_{kh} \geq 0, \quad \sigma_{kh} \in \{0, 1\}. \tag{29}$$

Conditions (25)–(29) express that, for each point (u_1, u_2), $\Lambda\big[f(u_1, u_2)\big]$ belongs to a convex set, in this specific case a tetrahedron determined by the following four points:

$$(a_k, b_h, f(a_k, b_h)), (a_{k+1}, b_h, f(a_{k+1}, b_h)),$$
$$(a_{k+1}, b_{h+1}, f(a_{k+1}, b_{h+1})), (a_k, b_{h+1}, f(a_k, b_{h+1})).$$

Let us remark that in order to guarantee that $\Lambda\big[f(u_1, u_2)\big]$ belongs to a convex planar subset (here: a triangle), conditions (25)–(29) could be reformulated. For this purpose, the rectangles (a_k, b_h), (a_{k+1}, b_h), (a_{k+1}, b_{h+1}), (a_k, b_{h+1}) are partitioned into two corresponding (inferior and superior) triangles, defined as (a_k, b_h), (a_{k+1}, b_h), (a_{k+1}, b_{h+1}) and (a_k, b_h), (a_{k+1}, b_{h+1}), (a_k, b_{h+1}), respectively. Then a straightforward reformulation of (25)–(29) may be obtained as follows:

$$\Lambda\big[f(u_1, u_2)\big] = \\ = \sum_{k=0}^{m-1}\sum_{h=0}^{p-1}\Big[(f(a_k, b_h)\lambda_{kh} + f(a_{k+1}, b_h)\widehat{\lambda}_{kh} + f(a_{k+1}, b_{h+1})\widehat{\widehat{\lambda}}_{kh}) \\ +(f(a_k, b_h)\mu_{kh} + f(a_{k+1}, b_{h+1})\widehat{\mu}_{kh} + f(a_k, b_{h+1})\widehat{\widehat{\mu}}_{kh})\Big], \tag{30}$$

$$u_1 = \sum_{k=0}^{m-1}\sum_{h=0}^{p-1}\Big[(a_k\lambda_{kh} + a_{k+1}\widehat{\lambda}_{kh} + a_{k+1}\widehat{\widehat{\lambda}}_{kh}) + (a_k\mu_{kh} + a_{k+1}\widehat{\mu}_{kh} + a_k\widehat{\widehat{\mu}}_{kh})\Big], \tag{31}$$

$$u_2 = \sum_{k=0}^{m-1}\sum_{h=0}^{p-1}\Big[(b_k\lambda_{kh} + b_k\widehat{\lambda}_{kh} + b_{k+1}\widehat{\widehat{\lambda}}_{kh}) + (b_k\mu_{kh} + b_{k+1}\widehat{\mu}_{kh} + b_{k+1}\widehat{\widehat{\mu}}_{kh})\Big], \tag{32}$$

$$\sum_{k=0}^{m-1}\sum_{h=0}^{p-1}(\lambda_{kh} + \widehat{\lambda}_{kh} + \widehat{\widehat{\lambda}}_{kh} + \mu_{kh} + \widehat{\mu}_{kh} + \widehat{\widehat{\mu}}_{kh}) = 1, \tag{33}$$

$$\forall k \in \{0, \dots, m-1\}, \forall h \in \{0, \dots, p-1\},$$
$$\lambda_{kh} + \widehat{\lambda}_{kh} + \widehat{\widehat{\lambda}}_{kh} = \sigma^-_{kh},\ \mu_{kh} + \widehat{\mu}_{kh} + \widehat{\widehat{\mu}}_{kh} = \sigma^+_{kh}. \tag{34}$$

Here, $\lambda_{kh}, \widehat{\lambda}_{kh}, \widehat{\widehat{\lambda}}_{kh}, \mu_{kh}, \widehat{\mu}_{kh}, \widehat{\widehat{\mu}}_{kh} \geq 0$, $\sigma^-_{kh} \in \{0, 1\}$, $\sigma^+_{kh} \in \{0, 1\}$.

Concerning conditions (25)–(29), and similarly (30)–(34), the number of binary variables σ_{kh} depends on the product mp corresponding to the breakpoints on the two axes (u_1, u_2) respectively. The variables σ_{kh} can be substituted with the binary variables $\sigma_{1k} \in \{0, 1\}$ and $\sigma_{2h} \in \{0, 1\}$, corresponding to $u_1 \in [a_0, a_m]$ and $u_2 \in [b_0, b_p]$, respectively (for $k \in \{0, \dots, m-1\}$, $h \in \{0, \dots, p-1\}$). Equalities (29) are thus replaced by the following constraints:

$$\forall k \in \{0, \dots, m-1\} \sum_{h=0}^{p-1}(\lambda_{kh} + \widehat{\lambda}_{kh} + \mu_{kh} + \widehat{\mu}_{kh}) = \sigma_{1k}, \tag{35}$$

$$\forall h \in \{0, \ldots, p-1\} \sum_{k=0}^{m-1} (\lambda_{kh} + \widehat{\lambda}_{kh} + \mu_{kh} + \widehat{\mu}_{kh}) = \sigma_{2h}, \tag{36}$$

$$\sum_{k=0}^{m-1} \sigma_{1k} = 1, \tag{37}$$

$$\sum_{h=0}^{p-1} \sigma_{2h} = 1, \tag{38}$$

$$\forall k \in \{0, \ldots, m-1\}, \forall h \in \{0, \ldots, p-1\}$$
$$\lambda_{kh} + \widehat{\lambda}_{kh} + \mu_{kh} + \widehat{\mu}_{kh} \geq \sigma_{1k} + \sigma_{2h} - 1. \tag{39}$$

Notice that inequalities (39) serve to tighten the MILP formulation. Further extensions concerning both non-rectangular (or non-triangular) discretization polytopes (cf. [14]) could be envisioned also for n-variate functions. Referring to formulation (25)–(29) instead of (35)–(39), for the sake of simplicity, appropriate variables $\sigma_{k_1 \ldots k_n} \in \{0, 1\}$ can be defined for this purpose. Nevertheless, increasing model dimensions could become a stumbling block in practice. If the resulting model is solved by means of a branch and bound strategy, then the extended equations corresponding to (29) could alternatively be eliminated and treated algorithmically, giving rise to generalized SOS2 based formulations. Regarding possible SOS2 extensions, we refer again to Williams [17].

As a further option, a heuristic approach, based on the generation of reduced models, may be conceived. For this purpose, subsets of variables $\sigma_{k_1 \ldots k_n}$ can be selected either randomly or by applying suitable deterministic criteria. The reduced model will contain only constraints derived from (29) that correspond to the selected variables $\sigma_{k_1 \ldots k_n}$.

6 Conclusions

The research presented here originates from a challenging control problem in space engineering: arguably, extensions are possible to other high-tech application areas such as robotics and automation. Here, we study the task of dispatching control for a dynamic system, through the available actuators: the control dispatch is optimized according to an overall performance criterion such as minimal total energy consumption, under stringent operational conditions. This control dispatch problem leads to a difficult nonlinear model with an objective function defined by non-convex separable univariate functions. In order to handle such model functions, PL approaches are frequently applied, in order to obtain approximate globally optimal solutions using a MILP formulation of the derived optimization problem. Several PL methods are available, ranging from specific algorithmic strategies to tailored modeling techniques. In this study, a comparison of three PL approaches has been carried out for the case study in question: these are the classical CL method, the so-called "disaggregated convex combination" (DCC) approach, and the method VN proposed by Vielma and Nemhauser. A review of these methods is presented, together with computational results. Our comparative analysis indicates that—within the specific application framework presented here—the DCC approach is competitive. Extensions of the basic DCC formulation have been also discussed. Our future research is aimed at broadening the experimental basis, and at addressing extensions towards multivariate functions and new applications.

Acknowledgements The authors thank Laureano Escudero for his interest and comments related to the present work.

References

1. Anselmi, A., Cesare, S., Dionisio, S., Fasano, G., Massotti, L.: Control propellant minimization for the next generation gravity mission. In: Fasano, G., Pintér, J.D. (eds.) Modeling and Optimization in Space Engineering – State of the Art and New Challenges. Springer, New York (2019)
2. Beale, E.M.L., Tomlin, J.A.: Special facilities in a general mathematical programming system for non-convex problems using ordered sets of variables. In: Lawrence, J. (ed.) Proceedings of the 5th International Conference on Operations Research, Tavistock, London (1969)
3. Beale, E.M.L., Forrest, J.J.H.: Global optimization using special ordered sets. Math. Program. **10**, 52–69 (1976)
4. Fasano, G.: Control dispatch in a spacecraft: an advanced optimization approach. In: 4th European Optimisation in Space Engineering (OSE) Workshop, 27–30 March 2017, University of Bremen (2017)
5. Fasano, G.: Dynamic system control dispatch: a global optimization approach. In: Fasano, G., Pintér, J.D. (eds.) Modeling and Optimization in Space Engineering – State of the Art and New Challenges. Springer, New York (2019)
6. Li, H.L., Yu, C.S.: Global optimization method for nonconvex separable programming problems. Eur. J. Oper. Res. **117**(2), 275–292 (1999)
7. Lin, M.H., Carlsson, J.G., Ge, D., Shi, J., Tsai, J.F.: A review of piecewise linearization methods. Math. Probl. Eng. (2013). http://dx.doi.org/10.1155/2013/101376
8. Misener, R.: ANTIGONE: algorithms for continuous/integer global optimization of nonlinear equations. J. Glob. Optim. **59**(2–3) (2014)
9. Nowak, I.: Relaxation and Decomposition Methods for Mixed Integer Nonlinear Programming. Birkhäuser Verlag (2005)
10. Pintér, J.D.: Global Optimization in Action. Kluwer Academic Publishers, Dordrecht (1996)
11. Pintér, J.D.: LGO—A Model Development and Solver System for Global-Local Nonlinear Optimization, User's Guide, Current edition (2016)
12. Taha, H.A.: Operations Research, 7th edn. Macmillan, New York, USA (2003)
13. Tawarmalani, M., Sahinidis, N.V.: Convexification and Global Optimization in Continuous and Mixed-Integer Nonlinear Programming: Theory, Algorithms, Software, and Applications. Kluwer Academic Publishers, Boston, MA (2002)
14. Vielma, J.P., Ahmed, S., Nemhauser, G.L.: Mixed-integer models for nonseparable piecewise linear optimization: unifying framework and extensions. Oper. Res. **58**, 303–315 (2010)

15. Vielma, J.P., Nemhauser, G.L.: Modeling disjunctive constraints with a logarithmic number of binary variables and constraints. Math. Program. **128**(1–2), 49–72 (2011)
16. Weisstein, E.W.: "Gray Code." From MathWorld—A Wolfram Web Resource. http://mathworld.wolfram.com/GrayCode.html. Accessed 13 June 2017
17. Williams, H.P.: Model Building in Mathematical Programming, 5th edn. Wiley, Chichester, West Sussex, United Kingdom (1990, 2013)

Smooth Minimization of Nonsmooth Functions with Parallel Coordinate Descent Methods

Olivier Fercoq and Peter Richtárik

Abstract We study the performance of a family of randomized parallel coordinate descent methods for minimizing a nonsmooth nonseparable convex function. The problem class includes as a special case L1-regularized L1 regression and the minimization of the exponential loss ("AdaBoost problem"). We assume that the input data defining the loss function is contained in a sparse $m \times n$ matrix A with at most ω nonzeros in each row and that the objective function has a "max structure", allowing us to smooth it. Our main contribution consists in identifying parameters with a closed-form expression that guarantees a parallelization speedup that depends on basic quantities of the problem (like its size and the number of processors). The theory relies on a fine study of the Lipschitz constant of the smoothed objective restricted to low dimensional subspaces and shows an increased acceleration for sparser problems.

Keywords Coordinate descent · Parallel computing · Smoothing · Lipschitz constant

1 Introduction

1.1 Motivation

More and more often, practitioners in machine learning, optimization, biology, engineering, and various industries need to solve optimization problems with a huge number of variables. In this setup, classical algorithms, which for historical reasons almost invariably focus on obtaining solutions of high accuracy, are not efficient

O. Fercoq (✉)
LTCI, Télécom ParisTech, Université Paris-Saclay, Paris, France
e-mail: olivier.fercoq@telecom-paristech.fr

P. Richtárik
School of Mathematics, The University of Edinburgh, Edinburgh, UK
e-mail: peter.richtarik@ed.ac.uk

J. D. Pintér and T. Terlaky (eds.), *Modeling and Optimization: Theory and Applications*, Springer Proceedings in Mathematics & Statistics 279,
https://doi.org/10.1007/978-3-030-12119-8_4

enough, or are outright unable to perform even a single iteration. Indeed, in the *big data optimization* setting, where the number N of variables is huge, inversion of matrices is not possible and even operations such as matrix vector multiplications are too expensive. Instead, attention is shifting towards simple methods with cheap iterations, low memory requirements, and good parallelization and scalability properties.

If the accuracy requirements are moderate and the problem has only simple constraints (such as box constraints), methods with these properties do exist: parallel coordinate descent methods [1, 26, 28, 35] emerged as a very promising class of algorithms in this domain. At each iteration, a *set* of coordinates is selected at random and the updates are processed and applied in parallel.

A major question when designing a Parallel Coordinate Descent Method (PCDM) is: how should we combine the updates computed by the various processors? One may simply compute the updates in the same way as in the single processor case, and apply them all. However, this strategy is doomed to fail: the method may end up oscillating between suboptimal points [35]. Indeed, although the individual updates are safe, there is no reason why adding them all up for should decrease the function value. In order to overcome this difficulty, Richtárik and Takáč [28] introduced the concept of Expected Separable Overapproximation (ESO). Thanks to this bound on the expected decrease after one iteration of the algorithm, they could define safe values for the amount of damping one should apply to the updates in order to have a converging algorithm.

They could prove a nearly linear theoretical parallelization speedup for a composite and partially separable functions. This means that the objective function is the sum of a separable nonsmooth function and a differentiable function of the form $f(x) = \sum_{J \in \mathcal{J}} f_J(x^{(J)})$ where each function f_J depends only on a small number of coordinates $x^{(J)}$. They also showed that the way coordinates are sampled has a huge impact on the performance. Indeed, PCDM implemented with a so-called τ-nice sampling can be faster than PCDM implemented with a more general uniform sampling by a factor $O(\sqrt{n})$, where n is the number of variables.

The goal of this paper is to study a class of nonsmooth functions on which similar parallelization speedups can be proved for parallel coordinate descent methods.

1.2 Brief Literature Review

Serial randomized methods. Leventhal and Lewis [12] studied the complexity of randomized coordinate descent methods for the minimization of convex quadratics and proved that the method converges linearly even in the non-strongly convex case. Linear convergence for smooth strongly convex functions was proved by Nesterov [20] and for general regularized problems by Richtárik and Takáč [27]. Complexity results for smooth problems with special regularizes (box constraints, L1 norm) were obtained by Shalev-Shwarz and Tewari [32] and Nesterov [20]. Nesterov was the first to analyze the block setting and proposed using different Lipschitz constants for different blocks. This has a big impact on the efficiency of the method since these constants capture important second-order information. Richtárik and Takáč [25, 27] improved, generalized, and simplified previous results and extended the analysis to the composite case. They gave the first analysis of a coordinate descent method using arbitrary probabilities. Lu and Xiao [14] recently studied the work developed in [20] and [28] and obtained further improvements. Coordinate descent methods were recently extended to deal with coupled constraints by Necoara et al. [22] and extended to the composite setting by Necoara and Patrascu [23]. When the function is not smooth neither composite, it is still possible to define coordinate descent methods with subgradients. An algorithm based on the averaging of past subgradient coordinates is presented in [37] and a successful subgradient-based coordinate descent method for problems with sparse subgradients is proposed by Nesterov [21]. Tappenden et al [39] analyzed an inexact randomized coordinate descent method in which proximal subproblems at each iteration are solved only approximately. Dang and Lan [5] studied complexity of stochastic block mirror descent methods for nonsmooth and stochastic optimization and an accelerated method was studies by Shalev-Shwarz and Zhang [33]. Lacoste-Julien et al. [11] were the first to develop a block coordinate Frank–Wolfe method. The generalized power method of Journée et al. [10] designed for sparse principal component analysis can be seen as a nonconvex block coordinate ascent method with two blocks [29].

Parallel methods. One of the first complexity results for a parallel coordinate descent method was obtained by Ruszczyński [30] and is known as the diagonal quadratic approximation method (DQAM). DQAM updates all blocks at each iteration, and hence is not randomized. The method was designed for solving a convex composite problem with quadratic smooth part and arbitrary separable nonsmooth part and was motivated by the need to solve separable linearly constrained problems arising in stochastic programming. As described in previous sections, a family of randomized parallel coordinate descent methods (PCDM) for convex composite problems was analyzed by Richtárik and Takáč [28]. Tappenden et al. [38] contrasted DQAM [30] with PCDM [28], improved the complexity result [28] in the strongly convex case, and showed that for PCDM, it is optimal choose τ to be equal to the number of processors. Improved complexity results for the weakly convex case were presented in [40]. Utilizing the ESO machinery [28] and the primal-dual technique developed by Shalev-Shwarz and Zhang [34], Takáč et al. [35] developed and analyzed a parallel (mini-batch) stochastic subgradient descent method (applied to the primal problem of training support vector machines with the hinge loss) and a parallel stochastic dual coordinate ascent method (applied to the dual box-constrained concave maximization problem). The analysis naturally extends to the general setting of Shalev-Shwarz and Zhang [34]. A parallel Newton coordinate descent method was proposed in [2]. Parallel methods for L1 regularized problems with an application to truss topology design were proposed by Richtárik and Takáč [26]. They give the first analysis of a greedy serial coordinate descent method for L1 regularized problems. An early analysis of a PCDM for L1 regularized problems was performed by Bradley et al. [1]. Other recent parallel methods include [15, 18].

Coordinate descent methods for nonsmooth functions. When the function is not smooth neither composite, it is still possible to define coordinate descent methods with subgradients. An algorithm based on the averaging of past subgradient coordinates is presented in [37] and a successful subgradient-based coordinate descent method is proposed in [21] for problems with sparse subgradients. In both cases, the convergence relies on the fact that, given a point x, for any coordinate i, one always chooses the directional derivative $g_i(x)$ corresponding to the same subgradient $g(x) \in \partial f(x)$.

In this paper, we follow an alternative approach, based on the smoothing technique introduced by Nesterov in [19]. If the function to optimize has a max structure, then one can define a smooth approximation of the function and minimize the approximation by any method available for smooth optimization, including coordinate descent. This work has been a motivation for the study of accelerated parallel coordinate descent [7] we did afterwards. Indeed, the smoothing technique is most useful with an accelerated algorithm, in order to get a $O(1/k)$ speed of convergence.

1.3 Contribution

- **Complexity results**. We give the first complexity results for minimizing functions of the type $F(x) = \max_{z \in Q}\{\langle Ax, z\rangle - g(z)\} + \sum_{i=1}^n \Psi_i(x^{(i)})$ by a parallel coordinate descent method. In fact, we are not aware of any complexity results even in the $\Psi \equiv 0$ case.
- **Nesterov separability**. We identify the number of nonzero entries in each row of the matrix A as the important quantity driving parallelization speedup.
- **ESO parameters**. We show that it is possible to compute Expected Separable Overapproximation parameters *easily*. It is very important as they are needed to run the parallel algorithm.
- **Complexity**. Our complexity results are spelled out in detail in Theorems 15 and 16, and are summarized in Table 1. The results are complete up to logarithmic factors and say that as long as SPCDM takes at least k iterations, where lower

Table 1 Summary of iteration complexity results

	Strong convexity	Convexity
Smooth problem with max structure		
Problem 6 (Theorem 15)	$\frac{n}{\tau} \times \frac{\frac{\beta(\tau)}{\mu\sigma}+\sigma_\Psi}{\sigma_{f_\mu}+\sigma_\Psi} \times \log(\frac{1}{\epsilon})$	$\frac{n\beta(\tau)}{\tau} \times \frac{2Diam^2}{\mu\sigma\epsilon} \times \log(\frac{1}{\epsilon})$
Nonsmooth problem with max structure		
Problem 7 (Theorem 16)	$\frac{n}{\tau} \times \frac{\frac{2\beta(\tau)D}{\epsilon\sigma}+\sigma_\Psi}{\sigma_{f_\mu}+\sigma_\Psi} \times \log(\frac{1}{\epsilon})$	$\frac{n\beta(\tau)}{\tau} \times \frac{8DDiam^2}{\sigma\epsilon^2} \times \log(\frac{1}{\epsilon})$

bounds for k are given in the table, then x_k is an ϵ-solution with probability at least $1-\rho$. The confidence level parameter ρ can't be found in the table as it appears in a logarithmic term which we suppressed from the table. More on the parameters: n is the number of blocks; τ is the number of processors; ϵ is the target accuracy; σ, μ and D are values related to the max structure that are defined in Sect. 2 of Sect. 2.2; σ_Ψ and σ_{f_μ} are the strong convexity parameters of the functions Ψ and f_μ; $Diam$ is the diameter of the level set of the loss function defined by the value of the loss function at the initial iterate x_0.

Observe that as τ increases, the number of iteration decreases. The actual rate of decrease is controlled by the value of $\beta(\tau)$ (as this is the only quantity that may grow with τ). In the convex case, any value of $\beta(\tau)$ smaller than τ leads to parallelization speedup. Under the "Nesterov separability assumption", we show that $\beta(\tau) \ll \tau$, leading to a nearly linear parallelization speedup.

- **Parallel randomized AdaBoost**. We observe that the *logarithm* of the exponential loss is of the form $f_\mu(x) = \log\left(\frac{1}{m}\sum_{j=1}^m \exp(b_j(Ax)_j)\right) = \max_{u\in\Delta_m}\sum_j(u_jb_j(Ax)_j - \mu u_j\log(u_j)) - \mu\log(m)$ with $\mu = 1$ and $\Delta_m = \{u : u \geq 0, \sum_i u_i = 1\}$. Our algorithm in this case can be interpreted as a parallel randomized boosting method [31]. More details are given in Sect. 6.3, and in a follow-up[1] paper [6]. Our complexity results improve on those in the machine learning literature. Moreover, our framework makes possible the use of regularizers. Note that Nesterov separability in the context of machine learning requires all examples to depend on at most ω features, which is often the case.
- **Subspace Lipschitz constants**. We derive simple formulas for Lipschitz constants of the gradient of f_μ associated with subspaces spanned by an arbitrary subset S of blocks (Sect. 3). As a special case, we show that the gradient of a Nesterov separable function is Lipschitz with respect to the separable norm $\|\cdot\|_{w^*}$ with constant equal to $\frac{\omega}{\upsilon\mu}$, where ω is degree of Nesterov separability. Besides being useful in our analysis, these results are also of independent interest in the design of gradient-based algorithms in big dimensions.

1.4 Contents

In Sect. 2, we describe the problems we study, the algorithm (smoothed parallel coordinate descent method) and review Nesterov's smoothing technique. In Sect. 3, we compute Lipschitz constants of the gradient of smooth approximations of Nesterov separable functions associated with subspaces spanned by arbitrary subset of blocks. In Sect. 4, we derive ESO inequalities. Complexity results are derived in Sect. 5 and finally, in Sect. 6, we describe three applications and preliminary numerical experiments.

---

[1] The results presented in this paper were obtained the Fall of 2012 and Spring of 2013, the follow-up work [6] was prepared in the Summer of 2013.

2 Smoothed Parallel Coordinate Descent Method

2.1 Nesterov's Smoothing Technique

Let $\mathbb{E}_1$ and $\mathbb{E}_2$ be two finite dimensional linear normed spaces, and $\mathbb{E}_1^*$ and $\mathbb{E}_2^*$ be their duals (i.e., the spaces of bounded linear functionals). We equip $\mathbb{E}_1$ and $\mathbb{E}_2$ with norms $\|\cdot\|_1$ and $\|\cdot\|_2$, and the dual spaces $\mathbb{E}_1^*$, $\mathbb{E}_2^*$ with the dual (conjugate norms):

$$\|y\|_j^* \stackrel{\text{def}}{=} \max_{\|x\|_j \leq 1} \langle y, x \rangle, \qquad y \in \mathbb{E}_j^*, \qquad j = 1, 2,$$

where $\langle y, x \rangle$ denotes the action of the linear functional y on x. Let $\bar{A} : \mathbb{E}_1 \to \mathbb{E}_2^*$ be a linear operator, and let $\bar{A}^* : \mathbb{E}_2 \to \mathbb{E}_1^*$ be its adjoint such that $\langle \bar{A}x, u \rangle = \langle x, \bar{A}^* u \rangle, \forall x \in \mathbb{E}_1, \forall u \in \mathbb{E}_2$.

Let us equip $\bar{A}$ with a norm as follows:

$$\begin{aligned}\|\bar{A}\|_{1,2} &\stackrel{\text{def}}{=} \max_{x,u} \left\{ \langle \bar{A}x, u \rangle \ : \ x \in \mathbb{E}_1, \ \|x\|_1 = 1, \ u \in \mathbb{E}_2, \ \|u\|_2 = 1 \right\} \\ &= \max_x \{ \|\bar{A}x\|_2^* \ : \ x \in \mathbb{E}_1, \ \|x\|_1 = 1 \} = \max_u \{ \|\bar{A}^* u\|_1^* \ : \ u \in \mathbb{E}_2, \ \|u\|_2 = 1 \}. \quad (1)\end{aligned}$$

Consider the function $\bar{f} : \mathbb{E}_1 \to \mathbb{R}$ given by

$$\bar{f}(x) = \max_{u \in \bar{Q}} \{ \langle \bar{A}x, u \rangle - \bar{g}(u) \},$$

where $\bar{Q} \subset \mathbb{E}_2$ is a compact convex set and $\bar{g} : \mathbb{E}_2 \to \mathbb{R}$ is convex. Clearly, $\bar{f}$ is convex and in general nonsmooth.

Let us describe Nesterov's smoothing technique for approximating $\bar{f}$ by a convex function with Lipschitz gradient. The technique relies on the introduction of a prox-function $\bar{d} : \mathbb{E}_2 \to \mathbb{R}$. This function is continuous and strongly convex on $\bar{Q}$ with convexity parameter $\bar{\sigma}$. Let u_0 be the minimizer of $\bar{d}$ on $\bar{Q}$. Without loss of generality, we can assume that $\bar{d}(u_0) = 0$ so that for all $u \in \bar{Q}$, $\bar{d}(u) \geq \frac{\bar{\sigma}}{2} \|u - u_0\|_2^2$. We also write $\bar{D} \stackrel{\text{def}}{=} \max\{ \bar{d}(u) \ : \ u \in \bar{Q} \}$. Nesterov's smooth approximation of $\bar{f}$ is defined for any $\mu > 0$ by

$$\bar{f}_\mu(x) \stackrel{\text{def}}{=} \max_{u \in \bar{Q}} \{ \langle \bar{A}x, u \rangle - \bar{g}(u) - \mu \bar{d}(u) \}. \quad (2)$$

Proposition 1 (Nesterov [19]) *The function $\bar{f}_\mu$ is continuously differentiable on $\mathbb{E}_1$ and satisfies*

$$\bar{f}_\mu(x) \leq \bar{f}(x) \leq \bar{f}_\mu(x) + \mu \bar{D}. \quad (3)$$

Moreover, $\bar{f}_\mu$ is convex and its gradient $\nabla \bar{f}_\mu(x) = \bar{A}^* u^*$, where u^* is the unique maximizer in (2)*, is Lipschitz continuous with constant $L_\mu = \frac{1}{\mu \bar{\sigma}} \|\bar{A}\|_{1,2}^2$. Hence, for all $x, h \in \mathbb{E}_1$,*

$$\bar{f}_\mu(x+h) \le \bar{f}_\mu(x) + \langle \nabla \bar{f}_\mu(x), h\rangle + \frac{\|\bar{A}\|_{1,2}^2}{2\mu\bar{\sigma}}\|h\|_1^2. \tag{4}$$

The above result will be used in this paper in various ways:

1. As a direct consequence of (4) for $\mathbb{E}_1 = \mathbb{R}^N$ (primal basic space), $\mathbb{E}_2 = \mathbb{R}^m$ (dual basic space), $\|\cdot\|_1 = \|\cdot\|_w$, $\|\cdot\|_2 = \|\cdot\|_v$, $\bar{d} = d$, $\bar{\sigma} = \sigma$, $\bar{Q} = Q$, $\bar{g} = g$, $\bar{A} = A$ and $\bar{f} = f$, we obtain the following inequality:

$$f_\mu(x+h) \le f_\mu(x) + \langle \nabla f_\mu(x), h\rangle + \frac{\|A\|_{w,v}^2}{2\mu\sigma}\|h\|_w^2. \tag{5}$$

2. A large part of this paper is devoted to various refinements (for a carefully chosen data-dependent w, we "replace" $\|A\|_{w,v}^2$ by an easily computable and interpretable quantity depending on h and ω, which gets smaller as h gets sparser and ω decreases) and extensions (left-hand side is replaced by $\mathbf{E}[f_\mu(x + h_{[\hat{S}]})]$) of inequality (5). In particular, we give formulas for fast computation of subspace Lipschitz constants of ∇f_μ and derive ESO inequalities—which are essential for proving iteration complexity results for variants of the smoothed parallel coordinate descent method introduced in this paper.
3. Besides the above application to smoothing f; we will utilize Proposition 1 also as a tool for computing Lipschitz constants of the gradient of two technical functions needed in proofs. In Sect. 3, we will use $\mathbb{E}_1 = \mathbb{R}^S$ ("primal update space" associated with a subset $S \subseteq [n]$), $\mathbb{E}_2 = \mathbb{R}^m$ and $\bar{A} = A^{(S)}$. In Sect. 4, we will use $\mathbb{E}_1 = \mathbb{R}^N$, $\mathbb{E}_2 = \mathbb{R}^{|\mathcal{P}|\times m}$ ("dual product space" associated with sampling $\hat{S}$) and $\bar{A} = \hat{A}$. These spaces and matrices will be defined in the abovementioned sections, where they are needed.

As shown in [19], the choice of the prox-function (and thus of the norm in E_2) can have a tremendous impact on the quality of the smoothing. For instance, let us consider

$$\bar{f}(x) = \max_{u\in\Sigma_m} \langle \bar{A}x, \bar{b}\rangle - \langle b, u\rangle,$$

where $\Sigma_m = \{u \in \mathbb{R}^m : \sum_{j=1}^m u^{(j)} = 1 \;\&\; u_j \ge 0,\ \forall j\}$ and the usual Euclidean distance (2-norm) in E_1. We shall compare prox-functions through the quantity $\bar{D}\|\bar{A}\|_{1,2}^2$ which appears in the complexity estimates of Theorems 15 and 16. If we choose the Euclidean prox-function in E_2 (1-strongly convex for the 2-norm, $p = 2$ in (10)), we get $\bar{D}\|\bar{A}\|_{1,2}^2 = \lambda_{\max}(A^T A)$. If we choose the entropy prox-function $d(u) = \log(m) + \sum_{j=1}^m u^{(j)} \log(u^{(j)})$, which is 1-strongly convex for the 1-norm on Σ_m ($p = 1$ in (10)), we get $\bar{D}\|\bar{A}\|_{1,2}^2 = \log(m) \max_{1\le j\le m} \sum_{i=1}^n A_{j,i}^2$. The norm of a row is usually much smaller than the spectral norm of the entire matrix, which implies that for this problem, one should choose the entropy as a prox-function. This basic example of function with a max structure shows that to smooth such functions properly, it is critical to be able to deal with a non-Euclidean setting; in particular with prox-functions that are strongly convex with respect to the 1-norm.

2.2 Nonsmooth and Smoothed Composite Problems

In this paper, we study the iteration complexity of PCDMs applied to two classes of convex composite optimization problems:

$$\text{minimize} \quad F(x) \stackrel{\text{def}}{=} f(x) + \Psi(x) \quad \text{subject to} \quad x \in \mathbb{R}^N, \tag{6}$$

$$\text{minimize} \quad F_\mu(x) \stackrel{\text{def}}{=} f_\mu(x) + \Psi(x) \quad \text{subject to} \quad x \in \mathbb{R}^N. \tag{7}$$

We assume (6) and (7) have an optimal solution (x^*) and consider the following setup.

1. **Structure of** f. We assume that f is of the form

$$f(x) \stackrel{\text{def}}{=} \max_{z \in Q}\{\langle Ax, z\rangle - g(z)\}, \tag{8}$$

   where $Q \subseteq \mathbb{R}^m$ is a nonempty compact convex set, $A \in \mathbb{R}^{m \times N}$, $g : \mathbb{R}^m \to \mathbb{R}$ is convex and $\langle \cdot, \cdot \rangle$ is the standard Euclidean inner product (the sum of products of the coordinates of the vectors). Note that f is convex and in general *nonsmooth*.
2. **Structure of** f_μ. We assume that f_μ is of the form

$$f_\mu(x) \stackrel{\text{def}}{=} \max_{z \in Q}\{\langle Ax, z\rangle - g(z) - \mu d(z)\}, \tag{9}$$

   where A, Q and g are as above, $\mu > 0$ and $d : \mathbb{R}^m \to \mathbb{R}$ is σ-strongly convex on Q with respect to the norm

$$\|z\|_v \stackrel{\text{def}}{=} \Big(\sum_{j=1}^m v_j^p |z_j|^p\Big)^{1/p}, \tag{10}$$

   where $v_1, \ldots, v_m$ are positive scalars, $1 \le p \le 2$ and $z = (z_1, \ldots, z_m)^T \in \mathbb{R}^m$. Further, we assume that d is nonnegative on Q and that $d(z_0) = 0$ for some $z_0 \in Q$. It follows that $d(z) \ge \frac{\sigma}{2}\|z - z_0\|_v^2$ for all $z \in Q$. Said otherwise, d is a prox-function on Q. We denote $D \stackrel{\text{def}}{=} \max_{z \in Q} d(z)$.
   For $p > 1$, let q be such that $\frac{1}{p} + \frac{1}{q} = 1$. The conjugate norm of $\|\cdot\|_v$ defined in (10) is given by

$$\|z\|_v^* \stackrel{\text{def}}{=} \max_{\|z'\|_v \le 1} \langle z', z\rangle = \begin{cases} \Big(\sum_{j=1}^m v_j^{-q}|z_j|^q\Big)^{1/q}, & 1 < p \le 2, \\ \max_{1 \le j \le m} v_j^{-1}|z_j|, & p = 1. \end{cases} \tag{11}$$

   By Proposition 1, f_μ is a *smooth* convex function: it is differentiable and its gradient is Lipschitz.

3. **Block structure**. Let $A = [A_1, A_2, \dots, A_n]$ be decomposed into nonzero column submatrices, where $A_i \in \mathbb{R}^{m \times N_i}$, $N_i \geq 1$ and $\sum_{i=1}^n N_i = N$, and $U = [U_1, U_2, \dots, U_n]$ be a decomposition of the $N \times N$ identity matrix U into submatrices $U_i \in \mathbb{R}^{N \times N_i}$. Note that $A_i = AU_i$. It will be useful to remark that

$$U_i^T U_j = \begin{cases} N_i \times N_i \text{ identity matrix}, & i = j, \\ N_i \times N_j \text{ zero matrix}, & \text{otherwise}. \end{cases} \tag{12}$$

For $x \in \mathbb{R}^N$, let $x^{(i)}$ be the block of variables corresponding to the columns of A captured by A_i, that is, $x^{(i)} = U_i^T x \in \mathbb{R}^{N_i}$, $i = 1, 2, \dots, n$. Clearly, any vector $x \in \mathbb{R}^N$ can be written uniquely as $x = \sum_{i=1}^n U_i x^{(i)}$. We will often refer to the vector $x^{(i)}$ as the *i-th block* of x. For $h \in \mathbb{R}^N$ and $\emptyset \neq S \subseteq [n] \stackrel{\text{def}}{=} \{1, 2, \dots, n\}$, it will be convenient to write

$$h_{[S]} \stackrel{\text{def}}{=} \sum_{i \in S} U_i h^{(i)}. \tag{13}$$

Finally, with each block i we associate a positive definite matrix $B_i \in \mathbb{R}^{N_i \times N_i}$ and scalar $w_i > 0$, and equip $\mathbb{R}^N$ with a pair of conjugate norms:

$$\|x\|_w^2 \stackrel{\text{def}}{=} \sum_{i=1}^n w_i \langle B_i x^{(i)}, x^{(i)} \rangle, \quad (\|y\|_w^*)^2 \stackrel{\text{def}}{=} \max_{\|x\|_w \leq 1} \langle y, x \rangle^2 = \sum_{i=1}^n w_i^{-1} \langle B_i^{-1} y^{(i)}, y^{(i)} \rangle. \tag{14}$$

Remark For some problems, it is relevant to consider blocks of coordinates as opposed to individual coordinates. The novel aspects of this paper are *not* in the block setup however, which was already considered in [20, 28]. We still write the paper in the general block setting; for several reasons. First, it is often practical to work with blocks either due to the nature of the problem (e.g., group lasso), or due to numerical considerations (it is often more efficient to process a "block" of coordinates at the same time). Second, some parts of the theory need to be treated differently in the block setting. The theory, however, does not get more complicated due to the introduction of blocks. A small notational overhead is a small price to pay for these benefits.

4. **Sparsity of** A. For a vector $x \in \mathbb{R}^N$ let

$$\Omega(x) \stackrel{\text{def}}{=} \{i \; : \; U_i^T x \neq 0\} = \{i \; : \; x^{(i)} \neq 0\}. \tag{15}$$

Let A_{ji} be the j-th row of A_i. If $e_1, \dots, e_m$ are the unit coordinate vectors in $\mathbb{R}^m$, then $A_{ji} \stackrel{\text{def}}{=} e_j^T A_i$. Using the above notation, the set of nonzero blocks of the j-th row of A can be expressed as

$$\Omega(A^T e_j) \stackrel{(15)}{=} \{i \; : \; U_i^T A^T e_j \neq 0\} = \{i \; : \; A_{ji} \neq 0\}. \tag{16}$$

The following concept is key to this paper.

Definition 2 (Nesterov separability)[2] We say that f (resp. f_μ) is *Nesterov (block) separable* of degree ω if it has the form (8) (resp. (9)) and

$$\max_{1\le j\le m} |\Omega(A^T e_j)| \le \omega. \tag{17}$$

Note that in the special case when all blocks are of cardinality 1 (i.e., $N_i = 1$ for all i), the above definition simply requires all rows of A to have at most ω nonzero entries.

5. **Separability of** Ψ. We assume that

$$\Psi(x) = \sum_{i=1}^{n} \Psi_i(x^{(i)}),$$

where $\Psi_i : \mathbb{R}^{N_i} \to \mathbb{R} \cup \{+\infty\}$ are *simple* proper closed convex functions.

Remark Note that we do not assume that the functions Ψ_i be smooth. In fact, the most interesting cases in terms of applications are nonsmooth functions such as, for instance, i) $\Psi_i(t) = \lambda|t|$ for some $\lambda > 0$ and all i (L1 regularized optimization), ii) $\Psi_i(t) = 0$ for $t \in [a_i, b_i]$, where $-\infty \le a_i \le b_i \le +\infty$ are some constants, and $\Psi_i(t) = +\infty$ for $t \notin [a_i, b_i]$ (box-constrained optimization).

2.3 The Algorithm

The method we use for solving the smoothed composite problem (7) is given in Algorithm 1.

SPCDM encodes a family of algorithms where each variant is characterized by the probability law governing $\hat{S}$. The sets generated throughout the iterations are assumed to be independent and identically distributed. In this paper, we focus on *uniform samplings*, which are characterized by the requirement that $\mathbf{P}(i \in \hat{S}) = \mathbf{P}(j \in \hat{S})$ for all $i, j \in [n]$. As a consequence, one has

$$\mathbf{P}(i \in \hat{S}) = \frac{\mathbf{E}[|\hat{S}|]}{n}. \tag{18}$$

In particular, we will focus on two special classes of uniform samplings: (i) those for which $\mathbf{P}(|\hat{S}| = \tau) = 1$ (τ-*uniform samplings*) and (ii) τ-uniform samplings with

[2] We coined the term *Nesterov separability* in honor of Yu. Nesterov's seminal work on the smoothing technique [19], which is applicable to functions represented in the form (8). Nesterov did not study problems with row-sparse matrices A, as we do in this work, nor did he study parallel coordinate descent methods. However, he proposed the celebrated smoothing technique which we also employ in this paper.

Algorithm 1 Smoothed Parallel Coordinate Descent Method (SPCDM)

Input: initial iterate $x_0 \in \mathbb{R}^N$, $\beta > 0$ and $w = (w_1, \dots, w_n) > 0$
for $k \geq 0$ **do**
Step 1. Generate a random set of blocks $S_k \subseteq \{1, 2, \dots, n\}$ following the law of $\hat{S}$
Step 2. In parallel for $i \in S_k$, compute

$$h_k^{(i)} = \arg\min_{t \in \mathbb{R}^{N_i}} \left\{ \langle (\nabla f_\mu(x_k))^{(i)}, t \rangle + \frac{\beta w_i}{2} \langle B_i t, t \rangle + \Psi_i(x_k^{(i)} + t) \right\}$$

Step 3. In parallel for $i \in S_k$, update $x_{k+1}^{(i)} \leftarrow x_k^{(i)} + h_k^{(i)}$ and set $x_{k+1}^{(j)} \leftarrow x_k^{(j)}$ for $j \notin S_k$
end for

the additional property that all subsets of cardinality τ are chosen equally likely (τ*-nice samplings*). We will also say that a sampling is *proper* if $\mathbf{P}(|\hat{S}| \geq 1) > 0$.

Let us remark that the scheme actually encodes an entire family of methods. For $\tau = 1$, we have a serial method (one block updated per iteration), for $\tau = n$, all the blocks are updated in each iteration, and there are many parallel methods in between, depending on the choice of τ.

On top of the random sampling, the algorithm depends on parameters $\beta > 0$ and $w \in \mathbb{R}^n_+$. These parameters are determined in such a way that the function f_μ satisfies an *Expected Separable Overapproximation* (ESO) defined for a function ϕ as

$$\mathbf{E}\left[\phi(x + h_{[\hat{S}]})\right] \leq \phi(x) + \frac{\mathbf{E}[|\hat{S}|]}{n} \left(\langle \nabla \phi(x), h \rangle + \frac{\beta}{2} \sum_{i=1}^{n} w_i \langle B_i h^{(i)}, h^{(i)} \rangle \right), \quad x, h \in \mathbb{R}^N, \tag{19}$$

where $h_{[\hat{S}]}$ is defined in (13). When (19) holds, we say that ϕ admits a (β, w)-ESO with respect to $\hat{S}$. For simplicity, we may sometimes write $(\phi, \hat{S}) \sim \mathrm{ESO}(\beta, w)$.

By doing so, we benefit from the following:

(i) We obtain a guaranteed upper bound on the function value at $x_{k+1} = x_k + (h_k)_{[S_k]}$.
(ii) Since the overapproximation is a convex quadratic in h, *it is easy to compute*

$$h_k = \arg\min_h f_\mu(x) + \frac{\mathbf{E}[|\hat{S}|]}{n} \left(\langle \nabla f_\mu(x), h \rangle + \frac{\beta}{2} \sum_{i=1}^{n} w_i \langle B_i h^{(i)}, h^{(i)} \rangle \right) + \mathbf{E}\left[\Psi(x + h_{[\hat{S}]})\right].$$

(iii) Since the overapproximation is block separable, one *can compute the updates* $h_k^{(i)}$ *in parallel* for all $i \in \{1, 2, \dots, n\}$: the formula is given in Step 2 of Algorithm 1.
(iv) Still from block separability, one *can compute the updates of* $i \in S_k$ *only*, where S_k is the sample set drawn at iteration k following the law describing $\hat{S}$.

Proving parallelization speedup. It is important to understand whether choosing $\tau > 1$ (several processors), as opposed to $\tau = 1$ (one single processor), leads to acceleration in terms of an improved complexity bound. By analogy with proximal gradient descent, we can see that $\frac{1}{\beta}$ can be interpreted as a stepsize. We would hence wish to choose small β, but not too small so that the method does not diverge. The issue of the computation of a good (small) parameter β is very intricate for several reasons and is at the heart of the design of a randomized parallel coordinate descent method. As can be seen from Table 1 in the introduction, the number of iterations required to obtain an ϵ-solution for Problem (6) or (7) is of the form $k \geq C(\epsilon)\frac{\beta(\tau)}{\tau}$, where $C(\epsilon)$ does not depend on τ. Hence, parallelization speedup occurs when the function

$$T(\tau) = \frac{\beta(\tau)}{\tau}$$

is decreasing. This has been proved for smooth partially separable functions in [28]. The goal of this paper is to prove it for Nesterov separable functions.

Cost of a single iteration. The arithmetic cost of a single iteration of SPCDM is $c = c_1 + c_2 + c_3$, where c_1 is the cost of computing the gradients $(\nabla f(x_k))^{(i)}$ for $i \in S_k$, c_2 is the cost of computing the updates $h_k^{(i)}$ for $i \in S_k$, and c_3 is the cost of applying these updates. For simplicity, assume that all blocks are of size 1 and that we update τ blocks at each iteration. Clearly, $c_3 = \tau$. Since often $h_k^{(i)}$ can be computed in closed form[3] and takes $O(1)$ operations, we have $c_2 = O(\tau)$. The value of c_1 is more difficult to predict in general since by Proposition 1, we have $\nabla f_\mu(x_k) = A^T z_k$, where $z_k = \arg\max_{z \in Q}\{\langle Ax_k, z\rangle - g(z) - \mu d(z)\}$, and hence c_1 depends on the relationship between A, Q, g and d. It is often the case though that z_{k+1} is obtained from z_k by changing at most δ coordinates, with δ being small. In such a case, it is efficient to maintain the vectors $\{z_k\}$ (update at each iteration will cost δ) and at iteration k to compute $(\nabla f_\mu(x_k))^{(i)} = (A^T z_k)^{(i)} = \langle a_i, z_k\rangle$ for $i \in S_k$, where a_i is the i-th column of A, whence $c_1 = \delta + 2\sum_{i \in S_k}\|a_i\|_0$. Since $\mathbf{P}(i \in S_k) = \tau/n$, we have

$$\mathbf{E}[c_1] = \delta + \tfrac{2\tau}{n}\sum_{i=1}^{n}\|a_i\|_0 = \delta + \tfrac{2\tau}{n}\,\mathrm{nnz}(A).$$

In summary, the expected overall arithmetic cost of a single iteration of SPCDM, under the assumptions made above, is $\mathbf{E}[c] = O(\frac{\tau}{n}\,\mathrm{nnz}(A) + \delta)$.

[3]This is the case in many cases, including (i) $\Psi_i(t) = \lambda_i|t|$, (ii) $\Psi_i(t) = \lambda_i t^2$, and (iii) $\Psi_i(t) = 0$ for $t \in [a_i, b_i]$ and $+\infty$ outside this interval (and the multivariate/block generalizations of these functions). For complicated functions $\Psi_i(t)$, one may need to do one-dimensional optimization, which will cost $O(1)$ for each i, provided that we are happy with an inexact solution. An analysis of PCDM in the $\tau = 1$ case in such an inexact setting can be found in Tappenden et al. [39], and can be extended to the parallel setting.

3 Fast Computation of Subspace Lipschitz Constants

Let us start by introducing the key concept of this section.

Definition 3 Let $\phi : \mathbb{R}^N \to \mathbb{R}$ be a smooth function and let $\emptyset \neq S \subseteq \{1, 2, \ldots, n\}$. Then we say that $L_S(\nabla\phi)$ is a *Lipschitz constant of* $\nabla\phi$ *associated with* S, with respect to norm $\|\cdot\|$, if

$$\phi(x + h_{[S]}) \leq \phi(x) + \langle \nabla\phi(x), h_{[S]}\rangle + \frac{L_S(\nabla\phi)}{2}\|h_{[S]}\|^2, \qquad x, h \in \mathbb{R}^N. \tag{20}$$

We will alternatively say that $L_S(\nabla\phi)$ is a subspace Lipschitz constant of $\nabla\phi$ *corresponding to the subspace spanned by blocks* i *for* $i \in S$, that is, $\{\sum_{i\in S} U_i x^{(i)} : x^{(i)} \in \mathbb{R}^{N_i}\}$, or simply a *subspace Lipschitz constant*.

Observe the above inequality can be equivalently written as

$$\phi(x + h) \leq \phi(x) + \langle \nabla\phi(x), h\rangle + \frac{L_{\Omega(h)}(\nabla\phi)}{2}\|h\|^2, \qquad x, h \in \mathbb{R}^N.$$

In this section, we will be concerned with obtaining easily computable formulas for subspace Lipschitz constants for $\phi = f_\mu$ with respect to the separable norm $\|\cdot\|_w$. Inequalities of this type were first introduced in [28] (therein called Deterministic Separable Overapproximation, or DSO). The basic idea is that in a parallel coordinate descent method in which τ blocks are updated at each iteration, subspace Lipschitz constants for sets S of cardinality τ are more relevant (and possibly much smaller = better) than the standard Lipschitz constant of the gradient, which corresponds to the special case $S = \{1, 2, \ldots, n\}$ in the above definition. This generalizes the concept of block/coordinate Lipschitz constants introduced by Nesterov [20] (in which case $|S| = 1$) to spaces spanned by multiple blocks.

We first derive a *generic* bound on subspace Lipschitz constants (Sect. 3.2), one that holds for any choice of w and v. Subsequently, we show (Sect. 3.3) that for a particular data-dependent choice of the parameters $w_1, \ldots, w_n > 0$ defining the norm in $\mathbb{R}^N$, the generic bound can be written in a very simple form from which it is clear that i) $L_S \leq L_{S'}$ whenever $S \subset S'$ and ii) that L_S decreases as the degree of Nesterov separability ω decreases. Moreover, it is important that the data-dependent weights w^* and the factor are easily computable, as these parameters are needed to run the algorithm.

3.1 Primal Update Spaces

As a first step, we need to construct a collection of normed spaces associated with the subsets of $\{1, 2, \ldots, n\}$. These will be needed in the technical proofs and also in the formulation of our results.

- **Spaces**. For $\emptyset \neq S \subseteq \{1, 2, \dots, n\}$, we define $\mathbb{R}^S \stackrel{\text{def}}{=} \bigotimes_{i \in S} \mathbb{R}^{N_i}$ and for $h \in \mathbb{R}^N$ we write $h^{(S)}$ for the vector in $\mathbb{R}^S$ obtained from h by deleting all coordinates belonging to blocks $i \notin S$ (and otherwise keeping the order of the coordinates).[4]
- **Matrices**. Likewise, let $A^{(S)} : \mathbb{R}^S \to \mathbb{R}^m$ be the matrix obtained from $A \in \mathbb{R}^{m \times N}$ by deleting all columns corresponding to blocks $i \notin S$, and note that

$$A^{(S)} h^{(S)} = A h_{[S]}. \tag{21}$$

- **Norms**. We fix positive scalars $w_1, w_2, \dots, w_n$ and on $\mathbb{R}^S$ define a pair of conjugate norms by

$$\|h^{(S)}\|_w \stackrel{\text{def}}{=} \left(\sum_{i \in S} w_i \langle B_i h^{(i)}, h^{(i)} \rangle \right)^{1/2}, \quad \|h^{(S)}\|_w^* \stackrel{\text{def}}{=} \left(\sum_{i \in S} w_i^{-1} \langle B_i^{-1} h^{(i)}, h^{(i)} \rangle \right)^{1/2}. \tag{22}$$

The standard Euclidean norm of a vector $h^{(S)} \in \mathbb{R}^S$ is given by

$$\|h^{(S)}\|_E^2 = \sum_{i \in S} \|h^{(i)}\|_E^2 = \sum_{i \in S} \langle h^{(i)}, h^{(i)} \rangle. \tag{23}$$

Remark Note that, in particular, for $S = \{i\}$ we get $h^{(S)} = h^{(i)} \in \mathbb{R}^{N_i}$ and $\mathbb{R}^S \equiv \mathbb{R}^{N_i}$ (primal block space); and for $S = [n]$ we get $h^{(S)} = h \in \mathbb{R}^N$ and $\mathbb{R}^S \equiv \mathbb{R}^N$ (primal basic space). Moreover, for all $\emptyset \neq S \subseteq [n]$ and $h \in \mathbb{R}^N$,

$$\|h^{(S)}\|_w = \|h_{[S]}\|_w, \tag{24}$$

where the first norm is in $\mathbb{R}^S$ and the second in $\mathbb{R}^N$.

3.2 General Bound

Our first result in this section, Theorem 4, is a refinement of inequality (5) for a sparse update vector h. The only change consists in the term $\|A\|_{w,v}^2$ being replaced by $\|A^{(S)}\|_{w,v}^2$, where $S = \Omega(h)$ and $A^{(S)}$ is the matrix, defined in Sect. 3.1, mapping vectors in the primal update space $\mathbb{E}_1 \equiv \mathbb{R}^S$ to vectors in the dual basic space $\mathbb{E}_2 \equiv \mathbb{R}^m$. The primal and dual norms are given by $\|\cdot\|_1 \equiv \|\cdot\|_w$ defined in (22) and $\|\cdot\|_2 \equiv \|\cdot\|_v$ defined in (10), respectively. This is indeed a refinement, since for any $\emptyset \neq S \subseteq [n]$,

[4]Note that $h^{(S)}$ is different from $h_{[S]} = \sum_{i \in S} U_i h^{(i)}$, which is a vector in $\mathbb{R}^N$, although both $h^{(S)}$ and $h_{[S]}$ are composed of blocks $h^{(i)}$ for $i \in S$.

$$\begin{aligned}\|A\|_{w,v} &\overset{(1)}{=} \max_{\substack{\|h\|_w=1\\ h\in\mathbb{R}^N}} \|Ah\|_v^* \\ &\geq \max_{\substack{\|h\|_w=1\\ h^{(i)}=0,\ i\in S\\ h\in\mathbb{R}^N}} \|Ah\|_v^* \overset{(13)}{=} \max_{\substack{\|h_{[S]}\|_w=1\\ h\in\mathbb{R}^N}} \|Ah_{[S]}\|_v^* \\ &\overset{(21)+(24)}{=} \max_{\substack{\|h^{(S)}\|_w=1\\ h\in\mathbb{R}^N}} \|A^{(S)}h^{(S)}\|_v^* \overset{(1)}{=} \|A^{(S)}\|_{w,v}.\end{aligned}$$

The improvement can be dramatic, and gets better for smaller sets S. Note that in the same manner, one can show that $\|A^{(S_1)}\|_{w,v} \leq \|A^{(S_2)}\|_{w,v}$ if $\emptyset \neq S_1 \subset S_2$.

Theorem 4 (Subspace Lipschitz Constants) *For any $x \in \mathbb{R}^N$ and nonzero $h \in \mathbb{R}^N$,*

$$f_\mu(x+h) \leq f_\mu(x) + \langle \nabla f_\mu(x), h\rangle + \frac{\|A^{(\Omega(h))}\|_{w,v}^2}{2\mu\sigma}\|h\|_w^2. \tag{25}$$

Proof. Fix $x \in \mathbb{R}^N$, $\emptyset \neq S \subseteq [n]$ and define $\bar{f} : \mathbb{R}^S \to \mathbb{R}$ by

$$\begin{aligned}\bar{f}(h^{(S)}) \overset{\text{def}}{=} f_\mu(x+h_{[S]}) &= \max_{u\in Q}\left\{\langle A(x+h_{[S]}), u\rangle - g(u) - \mu d(u)\right\} \\ &\overset{(21)}{=} \max_{u\in Q}\left\{\langle A^{(S)}h^{(S)}, u\rangle - \bar{g}(u) - \mu d(u)\right\},\end{aligned} \tag{26}$$

where $\bar{g}(u) = g(u) - \langle Ax, u\rangle$. Applying Proposition 1 (with $\mathbb{E}_1 = \mathbb{R}^S$, $\mathbb{E}_2 = \mathbb{R}^m$, $\bar{A} = A^{(S)}$, $\bar{Q} = Q$, $\|\cdot\|_1 = \|\cdot\|_w$ and $\|\cdot\|_2 = \|\cdot\|_v$), we conclude that the gradient of $\bar{f}$ is Lipschitz with respect to $\|\cdot\|_w$ on $\mathbb{R}^S$, with Lipschitz constant $\frac{1}{\mu\sigma}\|A^{(S)}\|_{w,v}^2$. Hence, for all $h \in \mathbb{R}^N$,

$$f_\mu(x+h_{[S]}) = \bar{f}(h^{(S)}) \leq \bar{f}(0) + \langle \nabla\bar{f}(0), h^{(S)}\rangle + \frac{\|A^{(S)}\|_{w,v}^2}{2\mu\sigma}\|h^{(S)}\|_w^2\ . \tag{27}$$

Note that $\nabla\bar{f}(0) = (A^{(S)})^T u^*$ and $\nabla f_\mu(x) = A^T u^*$, where u^* is the maximizer in (26), whence

$$\begin{aligned}\langle \nabla\bar{f}(0), h^{(S)}\rangle &= \langle (A^{(S)})^T u^*, h^{(S)}\rangle = \langle u^*, A^{(S)}h^{(S)}\rangle \overset{(21)}{=} \langle u^*, Ah_{[S]}\rangle \\ &= \langle A^T u^*, h_{[S]}\rangle = \langle \nabla f_\mu(x), h_{[S]}\rangle.\end{aligned}$$

Substituting this and the identities $\bar{f}(0) = f_\mu(x)$ and (24) into (27) gives

$$f_\mu(x+h_{[S]}) \leq f_\mu(x) + \langle \nabla f_\mu(x), h_{[S]}\rangle + \frac{\|A^{(S)}\|_{w,v}^2}{2\mu\sigma}\|h_{[S]}\|_w^2.$$

It just remains to observe that in view of (15) and (13), for all $h \in \mathbb{R}^N$ we have $h_{[\Omega(h)]} = h$. □

3.3 Bounds for Data-Dependent Weights w

From now on we will not consider arbitrary weight vector w but one defined by the data matrix A and the smoothing norm $\|z\|_v = (\sum_{j=1}^m v_j^p |z_j|^p)^{1/p}$ (10) as follows. Let us define $w^* = (w_1^*, \dots, w_n^*)$ by

$$w_i^* \overset{\text{def}}{=} \max\{(\|A_i B_i^{-1/2} t\|_v^*)^2 \;:\; t \in \mathbb{R}^{N_i},\; \|t\|_E = 1\}, \quad i = 1, 2, \dots, n. \tag{28}$$

If $N_i = 1$ and $B_i = 1$ for all i, we can simplify (28) into

$$w_i^* = \begin{cases} \max_{1\le j\le m} v_j^{-2} A_{ji}^2, & p = 1, \\ \left(\sum_{j=1}^m v_j^{-q} |A_{ji}|^q\right)^{2/q}, & 1 < p < 2, \\ \sum_{j=1}^m v_j^{-2} A_{ji}^2, & p = 2. \end{cases} \tag{29}$$

Notice that as long as the matrices $A_1, \dots, A_n$ are nonzero, we have $w_i^* > 0$ for all i, and hence the norm $\|\cdot\|_1 = \|\cdot\|_{w^*}$ is well defined. Letting $S = \{i\}$ and $\|\cdot\|_1 \equiv \|\cdot\|_{w^*}$, we see that w_i^* is defined so that $\|A^{(S)}\|_{w^*,v} = 1$. Indeed,

$$\begin{aligned} \|A^{(S)}\|_{w^*,v}^2 &\overset{(1)}{=} \max_{\|h^{(S)}\|_{w^*}=1} (\|A^{(S)} h^{(S)}\|_v^*)^2 \overset{(21)+(13)}{=} \max_{\|h^{(i)}\|_{w^*}=1} (\|AU_i h^{(i)}\|_v^*)^2 \\ &\overset{(14)+(22)}{=} \frac{1}{w_i^*} \max_{\|y^{(i)}\|_E=1} (\|AU_i B_i^{-1/2} y^{(i)}\|_v^*)^2 \overset{(28)}{=} 1. \end{aligned} \tag{30}$$

In the rest of this section, we establish an easily computable upper bound on $\|A^{(\Omega(h))}\|_{w^*,v}^2$ which will be useful in proving a complexity result for SPCDM used with a τ-uniform or τ-nice sampling.

The following is a technical lemma needed to establish the main result of this section.

Lemma 5 *For any $\emptyset \neq S \subseteq [n]$ and w^* chosen as in* (28), *the following hold:*

$$p = 1 \quad \Rightarrow \quad \max_{\|h^{(S)}\|_{w^*}=1} \max_{1\le j\le m} v_j^{-2} \sum_{i\in S} (A_{ji} h^{(i)})^2 \le 1,$$

$$1 < p \le 2 \quad \Rightarrow \quad \max_{\|h^{(S)}\|_{w^*}=1} \sum_{j=1}^m \left(v_j^{-q} \sum_{i\in S} (A_{ji} h^{(i)})^2 \right)^{q/2} \le 1.$$

Proof. For any $h^{(i)}$ define the transformed variable $y^{(i)} = (w_i^*)^{1/2} B_i^{1/2} h^{(i)}$ and note that

$$\|h^{(S)}\|_{w^*}^2 \overset{(22)+(14)}{=} \sum_{i\in S} w_i^* \langle B_i h^{(i)}, h^{(i)} \rangle = \sum_{i\in S} \langle y^{(i)}, y^{(i)} \rangle \overset{(23)}{=} \|y^{(S)}\|_E^2.$$

We will now prove the result separately for $p = 1$ and $1 < p \le 2$. For $p = 1$ we have

$$
\begin{aligned}
LHS &\overset{\text{def}}{=} \max_{\|h^{(S)}\|_{w^*}=1} \max_{1\le j\le m} v_j^{-2} \sum_{i\in S} (A_{ji}h^{(i)})^2 \\
&= \max_{\|y^{(S)}\|_E=1} \max_{1\le j\le m} \left(v_j^{-2} \sum_{i\in S} (w_i^*)^{-1} (A_{ji} B_i^{-1/2} y^{(i)})^2 \right) \\
&\le \max_{\|y^{(S)}\|_E=1} \left(\sum_{i\in S} (w_i^*)^{-1} \max_{1\le j\le m} v_j^{-2} (A_{ji} B_i^{-1/2} y^{(i)})^2 \right) \\
&= \max_{\|y^{(S)}\|_E=1} \left(\sum_{i\in S} \|y^{(i)}\|_E^2 \underbrace{(w_i^*)^{-1} \max_{1\le j\le m} v_j^{-2} \left(A_{ji} B_i^{-1/2} \tfrac{y^{(i)}}{\|y^{(i)}\|_E} \right)^2}_{\le \|A^{(\{i\})}\|^2_{w^*,v}=1} \right) \\
&\overset{(30)}{\le} \max_{\|y^{(S)}\|_E=1} \sum_{i\in S} \|y^{(i)}\|_E^2 \overset{(23)}{=} 1.
\end{aligned}
$$

For $1 < p \le 2$ we may write:

$$
\begin{aligned}
LHS &\overset{\text{def}}{=} \max_{\|h^{(S)}\|_{w^*}=1} \sum_{j=1}^m v_j^{-q} \left(\sum_{i\in S} (A_{ji}h^{(i)})^2 \right)^{q/2} \\
&= \max_{\|y^{(S)}\|_E=1} \sum_{j=1}^m v_j^{-q} \left(\sum_{i\in S} (w_i^*)^{-1} (A_{ji} B_i^{-1/2} y^{(i)})^2 \right)^{q/2}.
\end{aligned} \tag{31}
$$

We continue from (31), first by bounding $R \overset{\text{def}}{=} \sum_{i\in S} (w_i^*)^{-1} (A_{ji} B_i^{-1/2} y^{(i)})^2$ using the Hölder inequality in the form

$$
\sum_{i\in S} a_i b_i \le \left(\sum_{i\in S} |a_i|^s \right)^{1/s} \left(\sum_{i\in S} |b_i|^{s'} \right)^{1/s'},
$$

with $a_i = (w_i^*)^{-1} \left(A_{ji} B_i^{-1} \frac{y^{(i)}}{\|y^{(i)}\|_E} \right)^2 \|y^{(i)}\|^{2-2/s'}$, $b_i = \|y^{(i)}\|_E^{2/s'}$, $s = q/2$, and $s' = q/(q-2)$.

$$
\begin{aligned}
R^{q/2} &\le \left(\sum_{i\in S} (w_i^*)^{-q/2} \left| A_{ji} B_i^{-1} \tfrac{y^{(i)}}{\|y^{(i)}\|_E} \right|^q \|y^{(i)}\|_E^2 \right) \times \left(\sum_{i\in S} \|y^{(i)}\|_E^2 \right)^{(q-2)q/4} \\
&\le \sum_{i\in S} (w_i^*)^{-q/2} \left| A_{ji} B_i^{-1} \tfrac{y^{(i)}}{\|y^{(i)}\|_E} \right|^q \|y^{(i)}\|_E^2.
\end{aligned} \tag{32}
$$

where we used the fact that $\left(\sum_{i\in S} \|y^{(i)}\|_E^2\right)^{(q-2)q/4} \le 1$. We now substitute (32) into (31) and get:

$$
\begin{aligned}
LHS &\overset{(31)+(32)}{\leq} \max_{\|y^{(S)}\|_E=1} \left(\sum_{j=1}^m v_j^{-q} \sum_{i\in S} (w_i^*)^{-q/2} \left| A_{ji} B_i^{-1} \tfrac{y^{(i)}}{\|y^{(i)}\|_E} \right|^q \|y^{(i)}\|_E^2 \right)^{2/q} \\
&= \max_{\|y^{(S)}\|_E=1} \left(\sum_{i\in S} \|y^{(i)}\|_E^2 \underbrace{(w_i^*)^{-q/2} \sum_{j=1}^m v_j^{-q} \left| A_{ji} B_i^{-1} \tfrac{y^{(i)}}{\|y^{(i)}\|_E} \right|^q}_{\leq \left(\|A^{(\{i\})}\|^2_{w^*,v}\right)^{1/q} \leq 1} \right)^{2/q} \\
&\overset{(30)}{\leq} \max_{\|y^{(S)}\|_E=1} \left(\sum_{i\in S} \|y^{(i)}\|_E^2 \right)^{2/q} \\
&= \left(\max_{\|y^{(S)}\|_E=1} \sum_{i\in S} \|y^{(i)}\|_E^2 \right)^{2/q} \overset{(23)}{=} 1.
\end{aligned}
$$

□

Using the above lemma, we can now give a simple and easily interpretable bound on $\|A^{(S)}\|^2_{w^*,v}$.

Lemma 6 *For any* $\emptyset \neq S \subseteq [n]$ *and* w^* *chosen as in* (28),

$$\|A^{(S)}\|^2_{w^*,v} \leq \max_{1\leq j\leq m} |\Omega(A^T e_j) \cap S|.$$

Proof. It will be useful to note that

$$e_j^T A^{(S)} h^{(S)} \overset{(21)}{=} e_j^T A h_{[S]} \overset{(13)}{=} \sum_{i\in S} A_{ji} h^{(i)}. \tag{33}$$

We will make use the following form of the Cauchy–Schwarz inequality: for scalars a_i, $i \in Z$, we have $(\sum_{i\in Z} a_i)^2 \leq |Z| \sum_{i\in Z} a_i^2$. For $1 < p \leq 2$, we may write

$$
\begin{aligned}
\|A^{(S)}\|^2_{w^*,v} &\overset{(1)}{=} \max_{\|h^{(S)}\|_{w^*}\leq 1} (\|A^{(S)}h^{(S)}\|_v^*)^2 \overset{(11)}{=} \max_{\|h^{(S)}\|_{w^*}=1} \left(\sum_{j=1}^m v_j^{-q} \left| e_j^T A^{(S)} h^{(S)} \right|^q \right)^{1/q} \\
&\overset{(33)+(16)}{=} \max_{\|h^{(S)}\|_{w^*}=1} \left(\sum_{j=1}^m v_j^{-q} \left(\left| \sum_{i\in\Omega(A^T e_j)\cap S} A_{ji} h^{(i)} \right|^2 \right)^{q/2} \right)^{2/q} \\
&\overset{\text{(Cauchy–Schwarz)}}{\leq} \max_{\|h^{(S)}\|_{w^*}=1} \left(\sum_{j=1}^m v_j^{-q} \left(\left|\Omega(A^T e_j)\cap S\right| \sum_{i\in\Omega(A^T e_j)\cap S} (A_{ji} h^{(i)})^2 \right)^{q/2} \right)^{2/q} \\
&\leq \max_{1\leq j\leq m} |\Omega(A^T e_j)\cap S| \times \max_{\|h^{(S)}\|_{w^*}=1} \left(\sum_{j=1}^m v_j^{-q} \left(\sum_{i\in S} (A_{ji} h^{(i)})^2 \right)^{q/2} \right)^{2/q} \\
&\overset{\text{(Lemma 5)}}{\leq} \max_{1\leq j\leq m} |\Omega(A^T e_j)\cap S|.
\end{aligned}
$$

The proof for $p = 1$ follows the same argument. □

Below, we state and prove the main result of this section. It says that the interesting but somewhat non-informative quantity $\|A^{(\Omega(h))}\|_{w,v}^2$ appearing in Theorem 4 can for $w = w^*$ be bounded by a very natural and easily computable quantity capturing the interplay between the sparsity pattern of the rows of A and the sparsity pattern of h.

Theorem 7 (Subspace Lipschitz Constants for $w = w^*$) *For $S \subseteq \{1, 2, \ldots, n\}$ let*

$$L_S \stackrel{def}{=} \max_{1 \le j \le m} |\Omega(A^T e_j) \cap S|. \tag{34}$$

Then for all $x, h \in \mathbb{R}^N$,

$$f_\mu(x+h) \le f_\mu(x) + \langle \nabla f_\mu(x), h \rangle + \frac{L_{\Omega(h)}}{2\mu\sigma} \|h\|_{w^*}^2. \tag{35}$$

Proof. In view of Theorem 4, we only need to show that $\|A^{(\Omega(h))}\|_{w^*,v}^2 \le L_{\Omega(h)}$. This directly follows from Lemma 6. □

Let us now comment on the meaning of this theorem:

1. Note that $L_{\Omega(h)}$ depends on A and h through their *sparsity pattern* only. Furthermore, μ is a user chosen parameter and σ depends on d and the choice of the norm $\|\cdot\|_v$, which is independent of the data matrix A. Hence, the term $\frac{L_{\Omega(h)}}{\mu\sigma}$ is *independent* of the *values* of A and h. Dependence on A is entirely contained in the weight vector w^*, as defined in (28).
2. For each S we have $L_S \le \min\{\max_{1\le j\le m} |\Omega(A^T e_j)|, |S|\} = \min\{\omega, |S|\} \le \omega$, where ω is the degree of Nesterov separability of f.

    (a) By substituting the bound $L_S \le \omega$ into (35), we conclude that the gradient of f_μ is Lipschitz with respect to the norm $\|\cdot\|_{w^*}$, with Lipschitz constant equal to $\frac{\omega}{\mu\sigma}$.

    (b) By substituting $U_i h^{(i)}$ in place of h in (35), we observe that the gradient of f_μ is *block Lipschitz* with respect to the norm $\langle B_i \cdot, \cdot \rangle^{1/2}$, with Lipschitz constant corresponding to block i equal to $L_i = \frac{w_i^*}{\mu\sigma}$:

$$f_\mu(x + U_i h^{(i)}) \le f_\mu(x) + \langle \nabla f_\mu(x), U_i h^{(i)} \rangle + \frac{L_i}{2} \langle B_i h^{(i)}, h^{(i)} \rangle, \quad x \in \mathbb{R}^N, \; h^{(i)} \in \mathbb{R}^{N_i}.$$

3. In some sense, it is more natural to use the norm $\|\cdot\|_L^2$ instead of $\|\cdot\|_{w^*}^2$, where $L = (L_1, \ldots, L_n)$ are the block Lipschitz constants $L_i = \frac{w_i^*}{\mu\sigma}$ of ∇f_μ. If we do this, then although the situation is very different, inequality (35) is similar to the one given for partially separable smooth functions in [28]. Indeed, the weights defining the norm are in both cases equal to the block Lipschitz constants (of f in [28] and of f_μ here). Moreover, the leading term in [28] is structurally comparable to the leading term $L_{\Omega(h)}$. Indeed, it is equal to $\max_S |\Omega(h) \cap S|$, where the maximum is taken over the block domains S of the constituent functions $f_S(x)$ in the representation of f revealing partial separability: $f(x) = \sum_S f_S(x)$.

4 Expected Separable Overapproximation (ESO)

In this section, we compute parameters β and w yielding an ESO for the pair $(\phi, \hat{S})$, where $\phi = f_\mu$ and $\hat{S}$ is a proper uniform sampling. If inequality (19) holds, we will for simplicity write $(\phi, \hat{S}) \sim \mathrm{ESO}(\beta, w)$.

In Sect. 4.1, we establish a link between ESO for $(\phi, \hat{S})$ and Lipschitz continuity of the gradient of a certain collection of functions. This link will enable us to compute the ESO parameters β, w for the smooth approximation of a Nesterov separable function f_μ, needed both for running Algorithm 1 and for the complexity analysis. In Sect. 4.2, we define certain technical objects that will be needed for further analysis. In Sect. 4.3, we prove a first ESO result, computing β for any $w > 0$ and *any* proper uniform sampling. The formula for β involves the norm of a certain large matrix, and hence is not directly useful as β is needed for running the algorithm. Also, this formula does not explicitly exhibit dependence on ω. Hence, it is not immediately apparent that β will be smaller for smaller ω, as one would expect. Subsequently, in Sect. 4.4, we specialize this result to τ-uniform samplings and then further to the more-specialized τ-nice samplings. As in the previous section, in these special cases, we show that the choice $w = w^*$ leads to very simple closed-form expressions for β, allowing us to get direct insight into parallelization speedup.

4.1 ESO and Lipschitz Continuity

In order to obtain the ESO, we define the collection of functions $\hat{\phi}_x : \mathbb{R}^N \to \mathbb{R}$ for $x \in \mathbb{R}^N$ by

$$\hat{\phi}_x(h) \stackrel{\text{def}}{=} \mathbf{E}\left[\phi(x + h_{[\hat{S}]})\right]. \tag{36}$$

Let us first establish some basic connections between ϕ and $\hat{\phi}_x$.

Lemma 8 *Let $\hat{S}$ be any sampling and $\phi : \mathbb{R}^N \to \mathbb{R}$ any function. Then for all $x \in \mathbb{R}^N$*

(i) if ϕ is convex, so is $\hat{\phi}_x$,
(ii) $\hat{\phi}_x(0) = \phi(x)$,
(iii) if $\hat{S}$ is proper and uniform, and $\phi : \mathbb{R}^N \to \mathbb{R}$ is continuously differentiable, then

$$\nabla\hat{\phi}_x(0) = \frac{\mathbf{E}[|\hat{S}|]}{n}\nabla\phi(x).$$

Proof. Fix $x \in \mathbb{R}^N$. Note that $\hat{\phi}_x(h) = \mathbf{E}[\phi(x + h_{[\hat{S}]})] = \sum_{S\subseteq[n]} \mathbf{P}(\hat{S} = S)\phi(x + U_S h)$, where $U_S \stackrel{\text{def}}{=} \sum_{i\in S} U_i U_i^T$. As $\hat{\phi}_x$ is a convex combination of convex functions, it is convex, establishing (i). Property (ii) is trivial. Finally,

$$\nabla\hat{\phi}_x(0) = \mathbf{E}\left[\nabla\, \phi(x + h_{[\hat{S}]})\Big|_{h=0}\right] = \mathbf{E}\left[U_{\hat{S}}\nabla\phi(x)\right] = \mathbf{E}\left[U_{\hat{S}}\right]\nabla\phi(x) = \frac{\mathbf{E}[|\hat{S}|]}{n}\nabla\phi(x).$$

The last equality follows from the observation that $U_{\hat{S}}$ is an $N \times N$ binary diagonal matrix with ones in positions (v, v) for coordinates $v \in \{1, 2, \ldots, N\}$ belonging to blocks $i \in \hat{S}$ only, coupled with (18). □

Now, we establish a connection between ESO and a uniform bound in x on the Lipschitz constants of the gradient "at the origin" of the functions $\{\hat{\phi}_x,\ x \in \mathbb{R}^N\}$. The result will be used for the computation of the parameters of ESO for Nesterov separable functions.

Theorem 9 *Let $\hat{S}$ be proper and uniform, and $\phi : \mathbb{R}^N \to \mathbb{R}$ be continuously differentiable. Then the following statements are equivalent:*

(i) $(\phi, \hat{S}) \sim \mathrm{ESO}(\beta, w)$,
(ii) $\hat{\phi}_x(h) \le \hat{\phi}_x(0) + \langle \nabla \hat{\phi}_x(0), h\rangle + \frac{1}{2}\frac{\mathbf{E}[|\hat{S}|]\beta}{n}\|h\|_w^2, \qquad x, h \in \mathbb{R}^N.$

Proof. We only need to substitute (36) and Lemma 8(ii–iii) into inequality (ii) and compare the result with (19). □

4.2 *Dual Product Space*

We construct a linear space associated with a fixed block sampling $\hat{S}$, and several derived objects which will depend on the distribution of $\hat{S}$. These objects will be needed in the proof of Theorem 10 and in further text.

- **Space**. Let $\mathcal{P} \stackrel{\text{def}}{=} \{S \subseteq [n] \ : \ p_S > 0\}$, where $p_S \stackrel{\text{def}}{=} \mathbf{P}(\hat{S} = S)$. The *dual product space* associated with $\hat{S}$ is defined by

$$\mathbb{R}^{|\mathcal{P}|m} \stackrel{\text{def}}{=} \bigotimes_{S \in \mathcal{P}} \mathbb{R}^m.$$

- **Norms**. Letting $u = \{u^S \in \mathbb{R}^m \ : \ S \in \mathcal{P}\} \in \mathbb{R}^{|\mathcal{P}|m}$, we now define a pair of conjugate norms in $\mathbb{R}^{|\mathcal{P}|m}$ associated with v and $\hat{S}$:

$$\|u\|_{\hat{v}} \stackrel{\text{def}}{=} \Big(\sum_{S \in \mathcal{P}} p_S \|u^S\|_v^2\Big)^{1/2}, \qquad \|u\|_{\hat{v}}^* \stackrel{\text{def}}{=} \max_{\|u'\|_{\hat{v}} \le 1} \langle u', u\rangle = \Big(\sum_{S \in \mathcal{P}} p_S^{-1} (\|u^S\|_v^*)^2\Big)^{1/2}. \tag{37}$$

The notation $\hat{v}$ indicates dependence on both v and $\hat{S}$.
- **Matrices**. For each $S \in \mathcal{P}$ let

$$\hat{A}^S \stackrel{\text{def}}{=} p_S A \sum_{i \in S} U_i U_i^T \in \mathbb{R}^{m \times N}. \tag{38}$$

We define the matrix $\hat{A} \in \mathbb{R}^{|\mathcal{P}|m \times N}$, obtained by stacking the matrices $\hat{A}^S$, $S \in \mathcal{P}$, on top of each other (in the same order the vectors u^S, $S \in \mathcal{P}$ are stacked to form

$u \in \mathbb{R}^{|\mathcal{P}|m}$). The "hat" notation indicates that $\hat{A}$ depends on both A and $\hat{S}$. Note that $\hat{A}$ maps vectors from the primal basic space $\mathbb{E}_1 \equiv \mathbb{R}^N$ to vectors in the dual product space $\mathbb{E}_2 \equiv \mathbb{R}^{|\mathcal{P}|m}$. We use $\|\cdot\|_1 \equiv \|\cdot\|_w$ as the norm in $\mathbb{E}_1$ and $\|\cdot\|_2 \equiv \|\cdot\|_{\hat{v}}$ as the norm in $\mathbb{E}_2$. It will be useful to note that for $h \in \mathbb{R}^N$ and $S \in \mathcal{P}$, $(\hat{A}h)^S = \hat{A}^S h$.

4.3 Generic ESO for Proper Uniform Samplings

Our first ESO result covers all (proper) uniform samplings and is valid for any $w > 0$.

Theorem 10 (Generic ESO) *If $\hat{S}$ is proper and uniform, then*

$$(f_\mu, \hat{S}) \sim \mathrm{ESO}\left(\frac{n\|\hat{A}\|^2_{w,\hat{v}}}{\mu\sigma\mathbf{E}[|\hat{S}|]}, w\right) \tag{39}$$

Proof. Consider the function

$$\begin{aligned}\bar{f}(h) &\stackrel{\text{def}}{=} \mathbf{E}[f_\mu(x+h_{[\hat{S}]})] \stackrel{(2)}{=} \sum_{S\in\mathcal{P}} p_S \max_{u^S\in Q}\left\{\langle A(x+h_{[S]}), u^S\rangle - g(u^S) - \mu d(u^S)\right\}\\ &= \max_{\{u^S\in Q\,:\,S\in\mathcal{P}\}} \sum_{S\in\mathcal{P}} p_S\left\{\langle Ah_{[S]}, u^S\rangle + \langle Ax, u^S\rangle - g(u^S) - \mu d(u^S)\right\}. \end{aligned}\tag{40}$$

Let $u \in \bar{Q} \stackrel{\text{def}}{=} Q^{|\mathcal{P}|} \subseteq \mathbb{R}^{|\mathcal{P}|m}$ and note that $\sum_{S\in\mathcal{P}} p_S\langle Ah_{[S]}, u^S\rangle \stackrel{(38)+(13)}{=} \sum_{S\in\mathcal{P}}\langle \hat{A}^S h, u^S\rangle \stackrel{(38)}{=} \langle \hat{A}h, u\rangle$. We define $\bar{g} : \bar{Q} \to \mathbb{R}$ by $\bar{g}(u) \stackrel{\text{def}}{=} \sum_{S\in\mathcal{P}} p_S(g(u^S) - \langle Ax, u^S\rangle)$ and $\bar{d} : \bar{Q} \to \mathbb{R}$ by $\bar{d}(u) \stackrel{\text{def}}{=} \sum_{S\in\mathcal{P}} p_S d(u^S)$. Plugging all of the above into (40) gives $\bar{f}(h) = \max_{u\in\bar{Q}}\left\{\langle\hat{A}h, u\rangle - \bar{g}(u) - \mu\bar{d}(u)\right\}$. It is easy to see that $\bar{d}$ is σ-strongly convex on $\bar{Q}$ with respect to the norm $\|\cdot\|_{\hat{v}}$ defined in (37). Indeed, for any $u_1, u_2 \in \bar{Q}$ and $t \in (0,1)$,

$$\begin{aligned}\bar{d}(tu_1 + (1-t)u_2) &= \sum_{S\in\mathcal{P}} p_S d(tu_1^S + (1-t)u_2^S)\\ &\le \sum_{S\in\mathcal{P}} p_S\Big(td(u_1^S) + (1-t)d(u_2^S) - \frac{\sigma}{2}t(1-t)\|u_1^S - u_2^S\|^2_v\Big)\\ &\stackrel{(37)}{=} t\bar{d}(u_1) + (1-t)\bar{d}(u_2) - \frac{\sigma}{2}t(1-t)\|u_1 - u_2\|^2_{\hat{v}}.\end{aligned}$$

As $\bar{f}$ has a max structure, Proposition 1 (with $\mathbb{E}_1 = \mathbb{R}^N$, $\mathbb{E}_2 = \mathbb{R}^{|\mathcal{P}|m}$, $\bar{A} = \hat{A}$, $\|\cdot\|_1 = \|\cdot\|_w$, $\|\cdot\|_2 = \|\cdot\|_{\hat{v}}$ and $\bar{\sigma} = \sigma$) says that the gradient of $\bar{f}$ is Lipschitz with constant $\frac{1}{\mu\sigma}\|\hat{A}\|^2_{w,\hat{v}}$. We only need to apply Theorem 9, establishing (i). □

We now give an insightful characterization of $\|\hat{A}\|_{w,\hat{v}}$.

Theorem 11 *If $\hat{S}$ is proper and uniform, then*

$$\|\hat{A}\|^2_{w,\hat{v}} = \max_{h\in\mathbb{R}^N,\ \|h\|_w\le 1} \mathbf{E}\left[\left(\|Ah_{[\hat{S}]}\|^*_v\right)^2\right]. \tag{41}$$

Moreover, $\left(\frac{\mathbf{E}[|\hat{S}|]}{n}\right)^2 \|A\|^2_{w,v} \le \|\hat{A}\|^2_{w,\hat{v}} \le \min\left\{\mathbf{E}\left[\|A^{(\hat{S})}\|^2_{w,v}\right],\ \frac{\mathbf{E}[|\hat{S}|]}{n}\|A\|^2_{w,v},\ \max_{S\in\mathcal{P}} \|A^{(S)}\|^2_{w,v}\right\}$.

Proof. Identity (41) follows from

$$\begin{aligned}
\|\hat{A}\|_{w,\hat{v}} &\overset{(1)}{=} \max\{\langle \hat{A}h, u\rangle \ :\ \|h\|_w \le 1,\quad \|u\|_{\hat{v}} \le 1\}\\
&\overset{(37)}{=} \max\left\{\sum_{S\in\mathcal{P}} p_S\langle Ah_{[S]}, u^S\rangle \ :\ \|h\|_w \le 1,\quad \sum_{S\in\mathcal{P}} p_S\|u^S\|_v^2 \le 1\right\}. \qquad (42)\\
&= \max_{\|h\|_w\le 1}\max_{u}\left\{\sum_{S\in\mathcal{P}} p_S\|u^S\|_v\langle Ah_{[S]}, \tfrac{u^S}{\|u^S\|_v}\rangle \ :\ \sum_{S\in\mathcal{P}} p_S\|u^S\|_v^2 \le 1\right\}\\
&= \max_{\|h\|_w\le 1}\max_{\beta}\left\{\sum_{S\in\mathcal{P}} p_S\beta_S\|Ah_{[S]}\|^*_v \ :\ \sum_{S\in\mathcal{P}} p_S\beta_S^2 \le 1,\ \beta_S \ge 0\right\}\\
&= \max_{\|h\|_w\le 1}\left(\sum_{S\in\mathcal{P}} p_S\left(\|Ah_{[S]}\|^*_v\right)^2\right)^{1/2} = \max_{\|h\|_w\le 1}\left(\mathbf{E}\left[\left(\|Ah_{[\hat{S}]}\|^*_v\right)^2\right]\right)^{1/2}.
\end{aligned}$$

As a consequence, we now have

$$\begin{aligned}
\|\hat{A}\|^2_{w,\hat{v}} &\overset{(41)}{\le} \mathbf{E}\left[\max_{h\in\mathbb{R}^N,\ \|h\|_w\le 1}\left(\|Ah_{[\hat{S}]}\|^*_v\right)^2\right]\\
&\overset{(21)}{=} \mathbf{E}\left[\max_{h\in\mathbb{R}^N,\ \|h\|_w\le 1}\left(\|A^{(\hat{S})}h^{(\hat{S})}\|^*_v\right)^2\right] \overset{(1)}{=} \mathbf{E}\left[\|A^{(\hat{S})}\|^2_{w,v}\right] \le \max_{S\in\mathcal{P}}\|A^{(S)}\|^2_{w,v},
\end{aligned}$$

and $\|\hat{A}\|^2_{w,\hat{v}} \overset{(41)}{\le} \max_{h\in\mathbb{R}^N,\ \|h\|_w\le 1} \mathbf{E}\left[\|A\|^2_{w,v}\|h_{[\hat{S}]}\|^2_w\right] = \|A\|^2_{w,v}\max_{h\ \|h\|_w\le 1}\frac{\mathbf{E}[|\hat{S}|]}{n}\|h\|^2_w = \frac{\mathbf{E}[|\hat{S}|]}{n}\|A\|^2_{w,v}$.

Finally, restricting the vectors $\hat{u}^S$, $S\in\mathcal{P}$, to be equal (to z), we obtain the estimate

$$\begin{aligned}
\|\hat{A}\|_{w,\hat{v}} &\overset{(42)}{\ge} \max\{\mathbf{E}[\langle Ah_{[\hat{S}]}, z\rangle] \ :\ \|h\|_w \le 1,\quad \|z\|_v \le 1\}\\
&= \max\{\tfrac{\mathbf{E}[|\hat{S}|]}{n}\langle Ah, z\rangle \ :\ \|h\|_w \le 1,\quad \|z\|_v \le 1\} \overset{(1)}{=} \frac{\mathbf{E}[|\hat{S}|]}{n}\|A\|_{w,v},
\end{aligned}$$

giving the lower bound. □

In the following section, we will utilize ESO (39) for specific samplings. In particular, we give easily computable upper bounds on β in the special case when $w = w^*$.

4.4 ESO for Data-Dependent Weights w

Let us first establish ESO for τ-uniform samplings and $w = w^*$.

Theorem 12 (ESO for τ-uniform sampling) *If f is Nesterov separable of degree ω, $\hat{S}$ is a τ-uniform sampling, w^* is chosen as in (28), $\beta_1' \overset{def}{=} \min\{\omega, \tau\}$ and $\beta = \frac{\beta_1'}{\mu\sigma}$, then*

$$(f_\mu, \hat{S}) \sim \mathrm{ESO}\left(\beta, w^*\right) .$$

Proof. This follows from the ESO in Theorem 10 by using $\|\hat{A}\|_{w,v}^2 \overset{(41)}{\leq} \max_{S\in\mathcal{P}} \|A^{(S)}\|_{w,v}^2$ and $\|A^{(S)}\|_{w,v}^2 \leq \max_j |\Omega(A^T e_j) \cap S| \leq \min\{\omega, \tau\}$, $S \in \mathcal{P}$, which follows from Lemma 6 and the fact that $|\Omega(A^T e_j)| \leq \omega$ for all j and $|S| = \tau$ for all $S \in \mathcal{P}$. □

Before we establish an ESO result for τ-nice samplings, the main result of this section, we need a technical lemma with a number of useful relations. Identities (44) and (47) and estimate (48) are new, the other two identities are from [28]. For $S \subseteq [n] = \{1, 2, \ldots, n\}$ define

$$\chi_{(i\in S)} = \begin{cases} 1 & \text{if } i \in S, \\ 0 & \text{otherwise.} \end{cases} \tag{43}$$

Lemma 13 *Let $\hat{S}$ be any sampling, J_1, J_2 be nonempty subsets of $[n]$ and $\{\theta_{ij} : i \in [n],\ j \in [n]\}$ be any real constants. Then*

$$\mathbf{E}\left[\sum_{i\in J_1\cap\hat{S}} \sum_{j\in J_2\cap\hat{S}} \theta_{ij}\right] = \sum_{i\in J_1}\sum_{j\in J_2} \mathbf{P}(\{i, j\} \subseteq \hat{S})\theta_{ij}. \tag{44}$$

If $\hat{S}$ is τ-nice, then for any $\emptyset \neq J \subseteq [n]$, $\theta \in \mathbb{R}^n$ and $k \in [n]$, the following identities hold:

$$\mathbf{E}\left[\sum_{i\in J\cap\hat{S}} \theta_i \mid |J \cap \hat{S}| = k\right] = \frac{k}{|J|}\sum_{i\in J} \theta_i, \tag{45}$$

$$\mathbf{E}\left[|J \cap \hat{S}|^2\right] = \frac{|J|\tau}{n}\left(1 + \frac{(|J|-1)(\tau-1)}{\max(1, n-1)}\right), \tag{46}$$

$$\max_{1\leq i\leq n} \mathbf{E}[|J \cap \hat{S}| \times \chi_{(i\in\hat{S})}] = \frac{\tau}{n}\left(1 + \frac{(|J|-1)(\tau-1)}{\max(1, n-1)}\right). \tag{47}$$

Moreover, if $J_1, \ldots, J_m$ are subsets of $[n]$ of identical cardinality ($|J_j| = \omega$ for all j), then

$$\max_{1\le i\le n} \mathbf{E}[\max_{1\le j\le m} |J_j \cap \hat{S}| \times \chi_{(i\in\hat{S})}] \quad \le \quad \frac{\tau}{n}\sum_{k=1}^{k_{max}} \min\left\{1\ ,\ \frac{mn}{\tau}\sum_{l=\max\{k,k_{min}\}}^{k_{max}} c_l\pi_l\right\}, \tag{48}$$

where $k_{min} = \max\{1, \tau - (n-\omega)\}$, $k_{max} = \min\{\tau, \omega\}$, $c_l = \max\left\{\frac{l}{\omega}, \frac{\tau-l}{n-\omega}\right\} \le 1$ *if* $\omega < n$ *and* $c_l = \frac{l}{\omega} \le 1$ *otherwise, and*

$$\pi_l \stackrel{def}{=} \mathbf{P}(|J_j \cap \hat{S}| = l) = \frac{\binom{\omega}{k}\binom{n-\omega}{\tau-k}}{\binom{n}{\tau}}, \qquad k_{min} \le l \le k_{max}.$$

Proof. The first statement is a straightforward generalization of (26) in [28]. Identities (45) and (46) were established[5] in [28]. Let us prove (47). The statement is trivial for $n = 1$, assume therefore that $n \ge 2$. Notice that

$$\mathbf{E}[|J \cap \hat{S}| \times \chi_{(k\in\hat{S})}] = \mathbf{E}\left[\sum_{i\in J\cap\hat{S}}\ \sum_{j\in\{k\}\cap\hat{S}} 1\right] \stackrel{(44)}{=} \sum_{i\in J}\mathbf{P}(\{i,k\}\subseteq\hat{S}). \tag{49}$$

Using $\mathbf{P}(i \in \hat{S}) \stackrel{(18)}{=} \frac{\mathbf{E}[|\hat{S}|]}{n}$, and the simple fact that $\mathbf{P}(\{i,k\}\subseteq\hat{S}) = \frac{\tau(\tau-1)}{n(n-1)}$ whenever $i \neq k$, we get

$$\sum_{i\in J}\mathbf{P}(\{i,k\}\subseteq\hat{S}) = \begin{cases} \sum_{i\in J}\frac{\tau(\tau-1)}{n\max(1,n-1)} = \frac{|J|\tau(\tau-1)}{n(n-1)}, & \text{if } k \notin J,\\ \frac{\tau}{n} + \sum_{i\in J/\{k\}}\frac{\tau(\tau-1)}{n(n-1)} = \frac{\tau}{n}\left(1 + \frac{(|J|-1)(\tau-1)}{(n-1)}\right), & \text{if } k \in J. \end{cases} \tag{50}$$

Note that the expression in the $k \notin J$ case is smaller than expression in the $k \in J$ case. If we now combine (49) and (50) and take maximum in k, (47) is proved. Let us now establish (48). Fix i and let $\eta_j \stackrel{\text{def}}{=} |J_j \cap \hat{S}|$. We can estimate

$$\begin{aligned}
\mathbf{E}[\max_{1\le j\le m}\eta_j \times \chi_{(i\in\hat{S})}] &= \sum_{k=k_{\min}}^{k_{\max}} k\mathbf{P}\left(\max_{1\le j\le m}\eta_j \times \chi_{(i\in\hat{S})} = k\right)\\
&= \sum_{k=1}^{k_{\max}}\mathbf{P}\left(\max_{1\le j\le m}\eta_j \times \chi_{(i\in\hat{S})} \ge k\right)\\
&= \sum_{k=1}^{k_{\max}}\mathbf{P}\left(\bigcup_{j=1}^{m}\left\{\eta_j \ge k \ \&\ i\in\hat{S}\right\}\right)\\
&\le \sum_{k=1}^{k_{\max}}\min\left\{\mathbf{P}(i\in\hat{S}), \sum_{j=1}^{m}\mathbf{P}\left(\eta_j \ge k \ \ \&\ \ i\in\hat{S}\right)\right\}
\end{aligned}$$

---

[5] In fact, the proof of the former is essentially identical to the proof of (44), and (46) follows from (44) by choosing $J_1 = J_2 = J$ and $\theta_{ij} = 1$.

$$\overset{(18)}{=} \sum_{k=1}^{k_{\max}} \min \left\{ \frac{\tau}{n}, \sum_{j=1}^{m} \sum_{l=\max\{k,k_{\min}\}}^{k_{\max}} \mathbf{P}\left(\eta_j = l \;\&\; i \in \hat{S}\right) \right\}. \tag{51}$$

In the last step, we have used the fact that $\mathbf{P}(\eta_j = l) = 0$ for $l < k_{\min}$ to restrict the scope of l. Let us fix j and estimate $\mathbf{P}(\eta_j = l \;\&\; i \in \hat{S})$. Consider two cases:

(i) If $i \in J_j$, then among the $\binom{n}{\tau}$ equiprobable possible outcomes of the τ-nice sampling $\hat{S}$, the ones for which $|J_j \cap \hat{S}| = l$ and $i \in \hat{S}$ are those that select block i and $l-1$ other blocks from J_j ($\binom{\omega-1}{l-1}$ possible choices) and $\tau - l$ blocks from outside J_j ($\binom{n-\omega}{\tau-l}$ possible choices). Hence,

$$\mathbf{P}\left(\eta_j = l \;\;\&\;\; i \in \hat{S}\right) = \frac{\binom{\omega-1}{l-1}\binom{n-\omega}{\tau-l}}{\binom{n}{\tau}} = \frac{l}{\omega}\pi_l. \tag{52}$$

(ii) If $i \notin J_j$ (notice that this can not happen if $\omega = n$), then among the $\binom{n}{\tau}$ equiprobable possible outcomes of the τ-nice sampling $\hat{S}$, the ones for which $|\hat{S} \cap J_j| = l$ and $i \in \hat{S}$ are those that select block i and $\tau - l - 1$ other blocks from outside J_j ($\binom{n-\omega-1}{\tau-l-1}$ possible choices) and l blocks from J_j ($\binom{\omega}{l}$ possible choices). Hence,

$$\mathbf{P}\left(\eta_j = l \;\;\&\;\; i \in \hat{S}\right) = \frac{\binom{\omega}{l}\binom{n-\omega-1}{\tau-l-1}}{\binom{n}{\tau}} = \frac{\tau - l}{n - \omega}\pi_l. \tag{53}$$

It only remains to plug the maximum of (52) and (53) into (51). □

Theorem 14 (ESO for τ-nice sampling) *Let f be Nesterov separable of degree ω, $\hat{S}$ be τ-nice, and w^* be chosen as in* (28). *Then*

$$(f_\mu, \hat{S}) \sim \mathrm{ESO}(\beta, w^*),$$

where $\beta = \frac{\beta'}{\mu\sigma}$ and, if the dual norm $\|\cdot\|_v$ is defined with $p = 2$,

$$\beta' = \beta'_2 \overset{def}{=} 1 + \frac{(\omega-1)(\tau-1)}{\max(1, n-1)} \tag{54}$$

or, if $p = 1$,

$$\beta' = \beta'_3 \overset{def}{=} \sum_{k=1}^{k_{max}} \min \left\{ 1, \frac{mn}{\tau} \sum_{l=\max\{k,k_{min}\}}^{k_{max}} c_l \pi_l \right\} \tag{55}$$

where c_l, π_l, k_{min} *and* k_{max} *are as in Lemma 13:* $k_{min} = \max\{1, \tau - (n - \omega)\}$, $k_{max} = \min\{\tau, \omega\}$, $c_l = \max\left\{\frac{l}{\omega}, \frac{\tau - l}{n - \omega}\right\} \le 1$ *if* $\omega < n$, $c_l = \frac{l}{\omega} \le 1$ *otherwise, and* $\pi_l = \frac{\binom{\omega}{k}\binom{n-\omega}{\tau-k}}{\binom{n}{\tau}}$, $k_{min} \le l \le k_{max}$.

Proof. In view of Theorem 10, we only need to bound $\|\hat{A}\|^2_{w^*,\hat{v}}$. First, note that

$$\|\hat{A}\|^2_{w^*,\hat{v}} \overset{(1)}{=} \max_{\|h\|_{w^*}=1} (\|\hat{A}h\|^*_{\hat{v}})^2 \overset{(37)+(38)}{=} \max_{\|h\|_{w^*}=1} \sum_{S \in \mathcal{P}} p_S^{-1} (\|\hat{A}^S h\|^*_v)^2. \quad (56)$$

Further, it will be useful to observe that

$$\hat{A}^S_{ji} \overset{(38)}{=} p_S e_j^T A \sum_{k \in S} U_k U_k^T U_i \overset{(12)+(43)}{=} p_S \chi_{(i \in S)} A_{ji} \ . \quad (57)$$

For brevity, let us write $\eta_j \overset{\text{def}}{=} |\Omega(A^T e_j) \cap \hat{S}|$. As $\hat{S}$ is τ-nice, adding dummy dependencies if necessary, we can wlog assume that all rows of A have the same number of nonzero blocks: $|\Omega(A^T e_j)| = \omega$ for all j. Thus, $\pi_k \overset{\text{def}}{=} \mathbf{P}(\eta_j = k)$ does not depend on j. Consider now two cases, depending on whether the norm $\|\cdot\|_v$ in $\mathbb{R}^m$ is defined with $p = 1$ or $p = 2$.

(i) For $p = 2$, we can write

$$\begin{aligned}
\|\hat{A}\|^2_{w^*,\hat{v}} &\overset{(56)+(11)+(33)}{=} \max_{\|h\|_{w^*}=1} \sum_{S \in \mathcal{P}} p_S^{-1} \sum_{j=1}^m v_j^{-2} \left(\sum_{i=1}^n \hat{A}^S_{ji} h^{(i)} \right)^2 \\
&\overset{(57)}{=} \max_{\|h\|_{w^*}=1} \sum_{S \in \mathcal{P}} p_S^{-1} \sum_{j=1}^m v_j^{-2} \left(\sum_{i=1}^n p_S \chi_{(i \in S)} A_{ji} h^{(i)} \right)^2 \\
&\overset{(16)}{=} \max_{\|h\|_{w^*}=1} \sum_{S \in \mathcal{P}} p_S \sum_{j=1}^m v_j^{-2} \Big(\sum_{i \in \Omega(A^T e_j) \cap S} A_{ji} h^{(i)} \Big)^2 \\
&= \max_{\|h\|_{w^*}=1} \mathbf{E}\left[\sum_{j=1}^m v_j^{-2} \Big(\sum_{i \in \Omega(A^T e_j) \cap \hat{S}} A_{ji} h^{(i)} \Big)^2 \right] \\
&= \max_{\|h\|_{w^*}=1} \sum_{k=0}^n \mathbf{E}\left[\sum_{j=1}^m v_j^{-2} \Big(\sum_{i \in \Omega(A^T e_j) \cap \hat{S}} A_{ji} h^{(i)} \Big)^2 \,\Big|\, \eta_j = k \right] \pi_k \\
&= \max_{\|h\|_{w^*}=1} \sum_{k=0}^n \sum_{j=1}^m v_j^{-2} \mathbf{E}\left[\Big(\sum_{i \in \Omega(A^T e_j) \cap \hat{S}} A_{ji} h^{(i)} \Big)^2 \,\Big|\, \eta_j = k \right] \pi_k. \quad (58)
\end{aligned}$$

Using the Cauchy–Schwarz inequality, we can write

$$\mathbf{E}\Big[\Big(\sum_{i\in\Omega(A^Te_j)\cap\hat{S}} A_{ji}h^{(i)}\Big)^2\Big|\, \eta_j = k\Big] \overset{(CS)}{\leq} \mathbf{E}\Big[|\Omega(A^Te_j)\cap\hat{S}| \sum_{i\in\Omega(A^Te_j)\cap\hat{S}} (A_{ji}h^{(i)})^2\Big|\, \eta_j = k\Big]$$
$$= \mathbf{E}\Big[k \sum_{i\in\Omega(A^Te_j)\cap\hat{S}} (A_{ji}h^{(i)})^2\Big|\, \eta_j = k\Big]$$
$$\overset{(45)}{=} \frac{k^2}{\omega} \sum_{i\in\Omega(A^Te_j)} (A_{ji}h^{(i)})^2. \tag{59}$$

Combining (58) and (59), we finally get

$$\|\hat{A}\|^2_{w^*,\hat{v}} \leq \frac{1}{\omega}\sum_{k=0}^{n} k^2\pi_k \left(\max_{\|h\|_{w^*}=1} \sum_{j=1}^{m} v_j^{-2} \sum_{i=1}^{n} (A_{ji}h^{(i)})^2\right)$$
$$\overset{\text{(Lemma 5)}}{\leq} \frac{1}{\omega}\sum_{k=0}^{n} k^2\pi_k \overset{(46)}{=} \frac{\tau}{n}\left(1 + \frac{(\omega-1)(\tau-1)}{\max(1, n-1)}\right).$$

(ii) Consider now the case $p = 1$.

$$\|\hat{A}\|^2_{w^*,\hat{v}} \overset{(56)+(11)+(33)}{=} \max_{\|h\|_{w^*}=1} \sum_{S\in\mathcal{P}} p_S^{-1}\Big[\max_{1\leq j\leq m} v_j^{-2}\Big(\sum_{i=1}^{n} \hat{A}^S_{ji}h^{(i)}\Big)^2\Big]$$
$$\overset{(57)}{=} \max_{\|h\|_{w^*}=1} \sum_{S\in\mathcal{P}} p_S^{-1}\Big[\max_{1\leq j\leq m} v_j^{-2}\Big(\sum_{i=1}^{n} p_S \chi_{(i\in S)} A_{ji}h^{(i)}\Big)^2\Big]$$
$$\overset{(16)}{=} \max_{\|h\|_{w^*}=1} \sum_{S\in\mathcal{P}} p_S\Big[\max_{1\leq j\leq m} v_j^{-2}\Big(\sum_{i\in\Omega(A^Te_j)\cap S} A_{ji}h^{(i)}\Big)^2\Big]$$
$$\overset{\text{(Cauchy–Schwarz)}}{\leq} \max_{\|h\|_{w^*}=1} \sum_{S\in\mathcal{P}} p_S\Big[\max_{1\leq j\leq m} v_j^{-2}|\Omega(A^Te_j)\cap S| \sum_{i\in\Omega(A^Te_j)\cap S} (A_{ji}h^{(i)})^2\Big]$$
$$\leq \max_{\|h\|_{w^*}=1} \sum_{S\in\mathcal{P}} p_S\kappa_S\Big[\max_{1\leq j\leq m} v_j^{-2}\sum_{i\in S}(A_{ji}h^{(i)})^2\Big], \tag{60}$$

where $\kappa_S \overset{\text{def}}{=} \max_{1\leq j\leq m} |\Omega(A^Te_j)\cap S|$. Consider the change of variables $y^{(i)} = (w_i^*)^{1/2}B_i^{1/2}h^{(i)}$. Utilizing essentially the same argument as in the proof of Lemma 5 for $p = 1$, we obtain

$$\max_{1\leq j\leq m} v_j^{-2}\sum_{i\in S}(A_{ji}h^{(i)})^2 \leq \sum_{i\in S}\|y^{(i)}\|_E^2. \tag{61}$$

Since $\|y\|_E = \|h\|_{w^*}$, substituting (61) into (60) gives

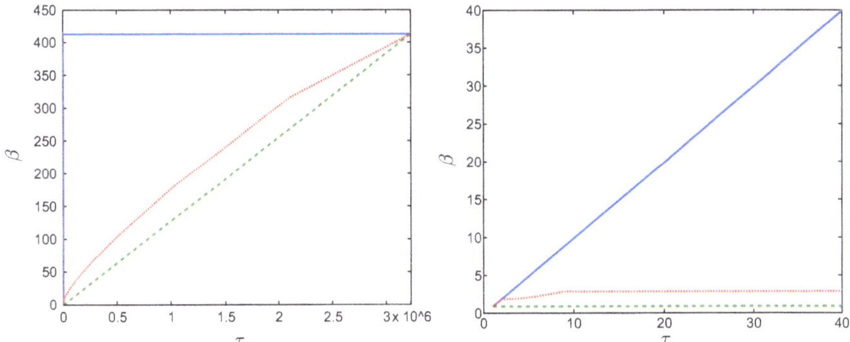

Fig. 1 Comparison of the three formulae for β' as a function of the number of processors τ (smaller β' is better). **Left**: large number of processors. **Right**: Zoom for smaller number of processors. We have used matrix $A \in \mathbb{R}^{m\times n}$ with $m = 2{,}396{,}130$, $n = 3{,}231{,}961$ and $\omega = 414$. **Blue solid line**: τ-uniform sampling, $\beta_1' = \min\{\omega, \tau\}$ (Theorem 12). **Green dashed line**: τ-nice sampling and $p = 2$, $\beta_2' = 1 + \frac{(\omega-1)(\tau-1)}{\max\{1,n-1\}}$ (Theorem 14). **Red dash-dotted line**: τ-nice sampling and $p = 1$, β_3' follows (55) in Theorem 14. Note that β_1' reaches its maximal value ω quickly, whereas β_2' increases slowly. When τ is small compared to n, this means that β_2' remains close to 1. As shown in Sect. 5 (see Theorems 16 and 15), small values of β' directly translate into better complexity and parallelization speedup

$$\begin{aligned}
\|\hat{A}\|^2_{w^*,\hat{v}} &\le \max_{\|y\|_E=1} \sum_{S\in\mathcal{P}} p_S\kappa_S \sum_{i\in S} \|y^{(i)}\|_E^2 = \max_{\|y\|_E=1} \sum_{i=1}^{n} \|y^{(i)}\|_E^2 \sum_{S\in\mathcal{P}} p_S\kappa_S\chi_{(i\in S)} \\
&= \max_{\|y\|_E=1} \sum_{i=1}^{n} \|y^{(i)}\|_E^2 \mathbf{E}[\kappa_{\hat{S}}\chi_{(i\in\hat{S})}] \\
&= \max_{1<i<n} \mathbf{E}[\kappa_{\hat{S}}\chi_{(i\in\hat{S})}] = \max_{1\le i\le n} \mathbf{E}\Big[\max_{1\le j\le m} |\Omega(A^T e_j)\cap\hat{S}| \times \chi_{(i\in\hat{S})}\Big]. \qquad (62)
\end{aligned}$$

It now only remains to apply inequality (48) used with $J_j = \Omega(A^T e_j)$. □

Let us now comment on some aspects of the above result.

1. Formula (55) may look complicated at first glance but it is in fact just a sum of a few easily computable terms. Computing β_3' has a negligible cost compared to the rest of the algorithm.
2. By comparing (47) and (48) one can see that $\beta_2' \le \beta_3' \le m\beta_2'$. However, in many situations, $\beta_3' \approx \beta_2'$ (see Fig. 1). Recall that a small β is good for Algorithm 1 (this will be formally proved in the next section).
3. If we let $\omega^* = \max_i\{j : A_{ji} \ne 0\}$ (maximum number of nonzero rows in matrices $A_1, \dots, A_n$), then in the $p = 1$ case we can replace β_3' by the smaller quantity

$$\beta_3'' \stackrel{\text{def}}{=} \frac{\tau}{n} \sum_{k=k_{\min}}^{k_{\max}} \min\left\{1, \sum_{l=k}^{k_{\max}} \left(m\frac{n}{n-\omega}\frac{\tau-l}{\tau} + n\frac{\omega^*}{\omega}\frac{l}{\tau}\right)\pi_l\right\}.$$

5 Iteration Complexity

In this section, we formulate *concrete* complexity results for Algorithm 1 applied to problem (6) by combining the generic results proved in [28] and outlined in the introduction, Proposition 1 (which draws a link between (6) and (7)) and, *most importantly*, the concrete values of β and w established in this paper for Nesterov separable functions and τ-uniform and τ-nice samplings.

A function $\phi : \mathbb{R}^N \to \mathbb{R} \cup \{+\infty\}$ is strongly convex with respect to the norm $\|\cdot\|_w$ with convexity parameter $\sigma_\phi(w) \geq 0$ if for all $x, \bar{x} \in \operatorname{dom} \phi$ and any subgradient $\phi'(\bar{x})$ of ϕ at $\bar{x}$,

$$\phi(x) \geq \phi(\bar{x}) + \langle \phi'(\bar{x}), x - \bar{x} \rangle + \frac{\sigma_\phi(w)}{2} \|x - \bar{x}\|_w^2 .$$

For $x_0 \in \mathbb{R}^N$ we let $\mathcal{L}_\mu^\delta(x_0) \stackrel{\text{def}}{=} \{x \ : \ F_\mu(x) \leq F_\mu(x_0) + \delta\}$ and define its diameter by

$$\mathcal{D}_{w,\mu}^\delta(x_0) \stackrel{\text{def}}{=} \max_{x,y}\{\|x - y\|_w \ : \ x, y \in \mathcal{L}_\mu^\delta(x_0)\} .$$

It will be useful to recall some basic notation from Sect. 2 that the theorems of this section will refer to: $F(x) = f(x) + \Psi(x)$, and $F_\mu(x) = f_\mu(x) + \Psi(x)$, with

$$f(x) = \max_{z \in Q}\{\langle Ax, z\rangle - g(z)\}, \quad f_\mu(x) = \max_{z \in Q}\{\langle Ax, z\rangle - g(z) - \mu d(z)\},$$

where d is strongly convex on Q wrt $\|\cdot\|_v$ with constant σ and $D = \max_{x \in Q} d(x)$. Recall also that $\|\cdot\|_v$ is a weighted p norm on $\mathbb{R}^m$, with weights $v_1, \dots, v_m > 0$. Also recall that $\|\cdot\|_w$ is a norm defined as a weighted quadratic mean of the block-norms $\langle B_i x^{(i)}, x^{(i)}\rangle^{1/2}$, with weights $w_1, \dots, w_n > 0$.

Theorem 15 (Complexity: smoothed composite problem (7)) *Pick* $x_0 \in \operatorname{dom} \Psi$ *and let* $\{x_k\}_{k \geq 0}$ *be the sequence of random iterates produced by the smoothed parallel descent method (Algorithm 1) with the following setup:*

(i) $\{S_k\}_{k \geq 0}$ *is an iid sequence of* τ*-uniform samplings, where* $\tau \in \{1, 2, \dots, n\}$*,*
(ii) $w = w^*$*, where* w^* *is defined in* (28)*,*
(iii) $\beta = \frac{\beta'}{\sigma\mu}$*, where* $\beta' = 1 + \frac{(\omega-1)(\tau-1)}{\max\{1, n-1\}}$ *if the samplings are* τ*-nice and* $p = 2$*,* β' *is given by* (55) *if the samplings are* τ*-nice and* $p = 1$*, and* $\beta' = \min\{\omega, \tau\}$ *if the samplings are not* τ*-nice (*ω *is the degree of Nesterov separability).*

Choose error tolerance $0 < \epsilon < F_\mu(x_0) - \min_x F_\mu(x)$*, confidence level* $0 < \rho < 1$ *and iteration counter* k *as follows:*

(i) if F_μ *is strongly convex with* $\sigma_{f_\mu}(w^*) + \sigma_\Psi(w^*) > 0$*, choose*

$$k \geq \frac{n}{\tau} \times \frac{\frac{\beta'}{\mu\sigma} + \sigma_\Psi(w^*)}{\sigma_{f_\mu}(w^*) + \sigma_\Psi(w^*)} \times \log\left(\frac{F_\mu(x_0) - \min_x F_\mu(x)}{\epsilon\rho}\right),$$

(ii) *otherwise additionally assume that*[6] $\epsilon < \frac{2n\beta}{\tau}$ *and that*[7] $\beta' = \min\{\omega, \tau\}$, *and choose*

$$k \geq \frac{n\beta'}{\tau} \times \frac{2(\mathcal{D}^0_{w^*,\mu}(x_0))^2}{\mu\sigma\epsilon} \times \log\left(\frac{F_\mu(x_0) - \min_x F_\mu(x)}{\epsilon\rho}\right).$$

Then

$$\mathbf{P}(F_\mu(x_k) - \min_x F_\mu(x) \leq \epsilon) \geq 1 - \rho.$$

Proof. This follows from the generic complexity bounds proved by Richtárik and Takáč [40, Theorem 19(ii) and Theorem 20] and Theorems 12 and 14 giving formulas for β' and w^* for which $(f_\mu, \hat{S}) \sim \mathrm{ESO}(\frac{\beta'}{\sigma\mu}, w^*)$. □

Now we consider solving the nonsmooth problem (6) by applying Algorithm 1 to its smooth approximation (7) for a specific value of the smoothing parameter μ.

Theorem 16 (Complexity: nonsmooth composite problem (6)) *Pick* $x_0 \in \mathrm{dom}\,\Psi$ *and let* $\{x_k\}_{k\geq 0}$ *be the sequence of random iterates produced by the smoothed parallel descent method (Algorithm 1) with the same setup as in Theorem 15, where* $\mu = \frac{\epsilon'}{2D}$ *and* $0 < \epsilon' < F(x_0) - \min_x F(x)$. *Further, choose confidence level* $0 < \rho < 1$ *and iteration counter as follows:*

(i) *if* F_μ *is strongly convex with* $\sigma_{f_\mu}(w^*) + \sigma_\Psi(w^*) > 0$, *choose*

$$k \geq \frac{n}{\tau} \times \frac{\frac{2\beta' D}{\sigma\epsilon'} + \sigma_\Psi(w^*)}{\sigma_{f_\mu}(w^*) + \sigma_\Psi(w^*)} \times \log\left(\frac{2(F(x_0) - \min_x F(x)) + \epsilon'}{\epsilon'\rho}\right),$$

(ii) *otherwise additionally assume that* $(\epsilon')^2 < \frac{8nD\beta'}{\sigma\tau}$ *and that* $\beta' = \min\{\omega, \tau\}$, *and choose*

$$k \geq \frac{n\beta'}{\tau} \times \frac{8D(\mathcal{D}^{\epsilon'/2}_{w^*,0}(x_0))^2}{\sigma(\epsilon')^2} \times \log\left(\frac{2(F(x_0) - \min_x F(x)) + \epsilon'}{\epsilon'\rho}\right).$$

Then

$$\mathbf{P}(F(x_k) - \min_x F(x) \leq \epsilon') \geq 1 - \rho.$$

Proof. We will apply Theorem 15 with $\epsilon = \frac{\epsilon'}{2}$ and $\mu = \frac{\epsilon'}{2D}$. From Proposition 1 (used with $\bar{A} = A$, $\bar{f} = f$, $\bar{Q} = Q$, $\bar{d} = d$, $\|\cdot\|_2 = \|\cdot\|_v$, $\bar{\sigma} = \sigma$, $\bar{D} = D$ and

[6]This assumption is not restrictive as $\beta' \geq 1, n \geq \tau$ and ϵ is usually small. However, it is technically needed.

[7]Without the assumption $\beta' = \min\{\omega, \tau\}$, the algorithm still converges but with a proved complexity in $O(1/(\epsilon\rho))$ instead of $O(1/\epsilon \log(1/(\epsilon\rho)))$ [40]. In our experiments, we have never encountered a problem with using the more efficient τ-nice sampling even in the non-strongly convex case. In fact, this weaker result may just be an artifact of the analysis.

$\bar{f}_\mu = f_\mu$), we get $f_\mu(y) \leq f(y) \leq f_\mu(y) + \mu D$, and adding $\Psi(y)$ to all terms leads to $F_\mu(y) \leq F(y) \leq F_\mu(y) + \mu D$, for all $y \in \operatorname{dom} \Psi$. Hence, $\min_y F_\mu(y) \leq \min_y F(y) \leq \min_y F_\mu(y) + \mu D$ and by difference, $F(x) - \min_y F(y) \leq F_\mu(x) - \min_y F_\mu(y) + 2\mu D$. Case (i) directly follows from this inequality and Theorem 15.

In case (ii), we additionally need to argue that: (a) $\epsilon < \frac{2n\beta}{\tau}$ and (b) $\mathcal{D}^0_{w^*,\mu}(x_0) \leq \mathcal{D}^{\epsilon'/2}_{w^*,0}(x_0)$. Note that (a) is equivalent to the assumption $(\epsilon')^2 < \frac{8nD\beta'}{\sigma\tau}$. Notice that as long as $F_\mu(x) \leq F_\mu(x_0)$, we have $F(x) \overset{(3)}{\leq} F_\mu(x) + \frac{\epsilon'}{2} \leq F_\mu(x_0) + \frac{\epsilon'}{2} \overset{(3)}{\leq} F(x_0) + \frac{\epsilon'}{2}$, and hence $\mathcal{L}^0_\mu(x_0) \subset \mathcal{L}^{\epsilon'/2}_0(x_0)$, which implies (b). □

Let us now briefly comment on the results.

1. If we choose the separable regularizer $\Psi(x) = \frac{\delta}{2}\|x\|^2_{w^*}$, then $\sigma_\Psi(w^*) = \delta$ and the strong convexity assumption is satisfied, irrespective of whether f_μ is strongly convex or not. A regularizer of this type is often chosen in machine learning applications.
2. Theorem 16 covers the problem $\min F(x)$ and hence we have on purpose formulated the results without any reference to the smoothed problem (with the exception of dependence on $\sigma_{f_\mu}(w^*)$ in case (i)). We traded a (very) minor loss in the quality of the results for a more direct formulation.
3. As the confidence level is inside a logarithm, it is easy to obtain a high probability result with this randomized algorithm. For problem (6) in the non-strongly convex case, iteration complexity is $O((\epsilon')^{-2}\log(1/\epsilon')$, which is comparable to other techniques available for the minimization of nonsmooth convex functions such as the subgradient method. In the strongly convex case the dependence is $O((\epsilon')^{-1})$. Note, however, that in many applications solutions only of moderate or low accuracy are required, and the focus is on the dependence on the number of processors τ instead. In this regard, our methods have excellent theoretical parallelization speedup properties.
4. It is clear from the complexity results that as more processors τ are used, the method requires fewer iterations, and the speedup gets higher for smaller values of ω (the degree of Nesterov separability of f). However, the situation is even better if the regularized Ψ is strongly convex—the degree of Nesterov separability then has a weaker effect on slowing down parallelization speedup.
5. For τ-nice samplings, β changes depending on p (the type of dual norm $\|\cdot\|_v$). However, σ changes also, as this is the strong convexity constant of the prox-function d with respect to the dual norm $\|\cdot\|_v$.

6 Computational Experiments

In this section, we consider the application of the smoothed parallel coordinate descent method (SPCDM) to three special problems and comment on some preliminary computational experiments. For simplicity, in the examples all blocks are of size 1 ($N_i = 1$ for all i) and $\Psi \equiv 0$.

In all tests, we used a shared-memory workstation with 32 Intel Xeon processors at 2.6 GHz and 128 GB RAM. We have implemented SPCDM using C++ and OpenMP.

We departed slightly from theory with an asynchronous[8] version of the algorithm in order to limit communication costs and approximated τ-nice sampling by a τ-independent sampling as in [28] (the latter is very easy to generate in parallel; and also results in a uniform sampling for which our method applies). Indeed, we may see the partial derivatives computed in an asynchronous context as approximations to what would have been obtained with synchronization at every iteration. As shown in [4, 13], this does not significantly hinder the convergence of first-order methods, especially when we require a solution with mild accuracy as in the present case. We experienced that the asynchronous algorithm performs about twice as fast as its synchronous counterpart in practice.

6.1 L-Infinity Regression/Linear Programming

Here, we consider the problem of minimizing the function

$$f(x) = \|\tilde{A}x - \tilde{b}\|_\infty = \max_{u \in Q}\{\langle Ax, u\rangle - \langle b, u\rangle\},$$

where $\tilde{A} \in \mathbb{R}^{m\times n}$, $\tilde{b} \in \mathbb{R}^m$, $A = \begin{bmatrix} \tilde{A} \\ -\tilde{A} \end{bmatrix} \in \mathbb{R}^{2m\times n}$, $b = \begin{bmatrix} \tilde{b} \\ -\tilde{b} \end{bmatrix} \in \mathbb{R}^{2m}$ and Q is the unit simplex in $\mathbb{R}^{2m}$ $Q \overset{\text{def}}{=} \{u_j \in \mathbb{R}^{2m} : \sum_j u_j = 1,\ u_j \geq 0\}$. We choose the dual norm $\|\cdot\|_v$ in $\mathbb{R}^{2m}$ with $p = 1$ and $v_j = 1$ for all j. Further, we choose the prox-function $d(u) = \log(2m) + \sum_{j=1}^{2m} u_j \log(u_j)$ with center $u_0 = (1, 1, \ldots, 1)/(2m)$. It can be shown that $\sigma = 1$ and $D = \log(2m)$. Moreover, we let all blocks be of size 1 ($N_i = 1$), choose $B_i = 1$ for all i in the definition of the primal norm and

$$w_i^* \overset{(29)}{=} \max_{1\leq j\leq 2m} A_{ji}^2 = \max_{1\leq j\leq m} \tilde{A}_{ji}^2 .$$

The smooth approximation of f is given by

$$f_\mu(x) = \mu \log\left(\frac{1}{2m}\sum_{j=1}^{2m} \exp\left(\frac{e_j^T Ax - b_j}{\mu}\right)\right) . \tag{63}$$

Experiment. In this experiment, we minimize f_μ utilizing τ-nice sampling and parameter β given by (55). We first compare SPCDM (Algorithm 1) with several other methods, see Table 2.

We perform a small-scale experiment so that we can solve the problem directly as a linear program with GLPK. The simplex method struggles to progress initially

[8] Some synchronization do take place from time to time for monitoring purposes.

Table 2 Comparison of various algorithms for the minimization of $f(x) = \|\tilde{A}x - \tilde{b}\|_\infty$, where $\tilde{A}$ and $\tilde{b}$ are taken from the Dorothea dataset [9] ($m = 800$, $n = 100{,}000$, $\omega = 6{,}061$) and $\epsilon = 0.01$

Algorithm	# iterations	Time (s)
GLPK's simplex	55,899	681
Accelerated gradient [19], $\tau = 16$ cores	8,563	246
Sparse subgradient [21], optimal value known	1,730	6.4
Sparse subgradient [21], optimal value unknown	166,686	544
Smoothed serial coordinate descent with line search [20], $\tau = 1$	4,800,000	25
Smoothed PCDM (Theorem 15), $\tau = 1$ cores ($\beta = 1.0$)	25,100,000	33
Smoothed PCDM (Theorem 15), $\tau = 4$ cores ($\beta = 3.4$)	8,700,000	29
Smoothed PCDM (Theorem 15), $\tau = 16$ cores ($\beta = 6.3$)	4,200,000	22
Smoothed PCDM (Theorem 15), $\tau = 24$ cores ($\beta = 7.6$)	3,300,000	20

but eventually finds the exact solution quickly. The accelerated gradient algorithm of Nesterov is easily parallelizable, which makes it competitive, but it suffers from small stepsizes (we chose here the estimate for the Lipschitz constant of the gradient given in [21] for this problem). A very efficient algorithm for the minimization of the infinity norm is Nesterov's sparse subgradient method [21] that is the fastest in our tests even when it uses a single core only. It performs full subgradient iterations in a very cheap way, utilizing the fact that the subgradients are sparse. The method has a sublinear in n complexity. However, in order for the method to take long steps, one needs to know the optimal value in advance. Otherwise, the algorithm is much slower, as is shown in the table.

For this problem, and without the knowledge of the optimal value, the smoothed parallel coordinate descent method presented in this paper is the fastest algorithm. Many iterations are needed but they are very cheap: in its serial version, at each iteration one only needs to compute one partial derivative and to update 1 coordinate of the optimization variable, the residuals and the normalization factor. The worst case algorithmic complexity of one iteration is thus proportional to the number of nonzero elements in one column; on average.

Observe that quadrupling the number of cores does not divide by 4 the computational time because of the increase in the β parameter. Yet, a non negligible speedup is observed when increasing the computational power.

Another possibility is to employ a line search strategy (see [20] for a line search adapted to coordinate descent), in order to take better advantage of the local curvature of the function. On our problem, this leads to a 25% speedup. However, designing an efficient line search procedure for *parallel* coordinate descent is still the subject of current research. A critical aspect would be to perform the line search independently along each coordinate.

Remark There are numerical issues with the smooth approximation of the infinity norm because it involves exponentials of potentially large numbers. A safe way of computing (63) is to compute first $\bar{r} = \max_{1 \leq j \leq 2m}(Ax - b)_j$ and to use the formula

$$f_\mu(x) = \bar{r} + \mu \log\left(\frac{1}{2m}\sum_{j=1}^{2m} \exp\left(\frac{(Ax-b)_j - \bar{r}}{\mu}\right)\right).$$

However, this formula is not suitable for parallel updates because the logarithm prevents us from making reductions. We adapted it in the following way to deal with parallel updates. Suppose we have already computed $f_\mu(x)$. Then $f_\mu(x+h) = f_\mu(x) + \mu \log(S_x(\delta))$, where

$$S_x(h) \stackrel{\text{def}}{=} \frac{1}{2m}\sum_{j=1}^{2m} \exp\left(\frac{(Ax-b)_j + (Ah)_j - f_\mu(x)}{\mu}\right)$$

In particular, $S_x(0) = 1$. Thus, as long as the updates are reasonably small, one can compute the exponentially growing quantity $\exp[((Ax-b)_j + (Ah)_j - f_\mu(x))/\mu]$ and update the sum in parallel. From time to time (for instance every n iterations or when S_x becomes small), we recompute $f_\mu(x)$ from scratch and reset h to zero.

6.2 L1 Regression

Here we consider the problem of minimizing the function

$$f(x) = \|Ax - b\|_1 = \max_{u \in Q}\{\langle Ax, u\rangle - \langle b, u\rangle\},$$

where $Q = [-1, 1]^n$. We define the dual norm $\|\cdot\|_v$ with $p = 2$ and $v_j = \sum_{i=1}^n A_{ji}^2$ for all $j = 1, 2, \ldots, m$. Further, we choose the prox-function $d(z) = \frac{1}{2}\|z\|_v^2$ with center $z_0 = 0$. Clearly, $\sigma = 1$ and $D = \frac{1}{2}\sum_{j=1}^m v_j = \sum_{j=1}^m \sum_{i=1}^n A_{ji}^2 = \|A\|_F^2$. Moreover, we choose $B_i = 1$ for all $i = 1, 2, \dots, n$ in the definition of the primal norm and $w_i^* \overset{(29)}{=} \sum_{j=1}^m v_j^{-2} A_{ji}^2$, $\forall i \in \{1, 2, \dots, n\}$.

The smooth approximation of f is given by

$$f_\mu(x) = \sum_{j=1}^m \|e_j^T A\|_{w^*}^* \psi_\mu \left(\frac{|e_j^T Ax - b_j|}{\|e_j^T A\|_v^*} \right), \qquad \psi_\mu(t) = \begin{cases} \frac{t^2}{2\mu}, & 0 \le t \le \mu, \\ t - \frac{\mu}{2}, & \mu \le t. \end{cases}$$

Remark Note that in [19], the dual norm is defined from the primal norm. In the present work, we need to define the dual norm first since otherwise the definitions of the norms would cycle. However, the definitions above give the choice of v that minimizes the term $D\|e\|_{w^*}^2 = D \sum_{i=1}^n w_i^* = (\sum_{j=1}^m v_j)(\sum_{j'=1}^m v_{j'}^{-2} A_{j'i}^2)$, where $e = (1, 1, \dots, 1) \in \mathbb{R}^N$. We believe that in the non-strongly convex case, one can replace in the complexity estimates the squared diameter of the level set by $\|x_0 - x^*\|_{w^*}^2$, which would then mean that a product of the form $D\|x_0 - x_*\|_{w^*}^2$ appears in the complexity. The above choice of the weights $v_1, \dots, v_m$ minimizes this product under assuming that $x_0 - x^*$ is proportional to e.

Experiment. We performed our medium scale numerical experiments (in the case of L1 regression and exponential loss minimization (Sect. 6.3)) on the URL reputation dataset [17]. It gathers $n = 3{,}231{,}961$ features about $m = 4{,}792{,}260$ URLs collected during 120 days. The feature matrix is sparse but it has some dense columns. The maximum number of nonzero elements in a row is $\omega = 414$. The vector of labels classifies the page as spam or not.

We applied SPCDM with τ-nice sampling, following the setup described in Theorem 16. The results for $f(x) = \|Ax - b\|_1$ are gathered in Fig. 2. We can see that parallelization speedup is proportional to the number of processors. In the right plot, we observe that the algorithm is not monotonic but monotonic on average.

6.3 Logarithm of the Exponential Loss

Here, we consider the problem of minimizing

$$f_1(x) = \log \left(\frac{1}{m} \sum_{j=1}^m \exp(b_j (Ax)_j) \right). \tag{64}$$

The AdaBoost algorithm [8] minimizes the exponential loss $\exp(f_1(x))$ by a greedy serial coordinate descent method (i.e., at each iteration, one selects the coordinate corresponding to the largest directional derivative and updates that coordinate only). We observe that f_1 is Nesterov separable as it is the smooth approximation

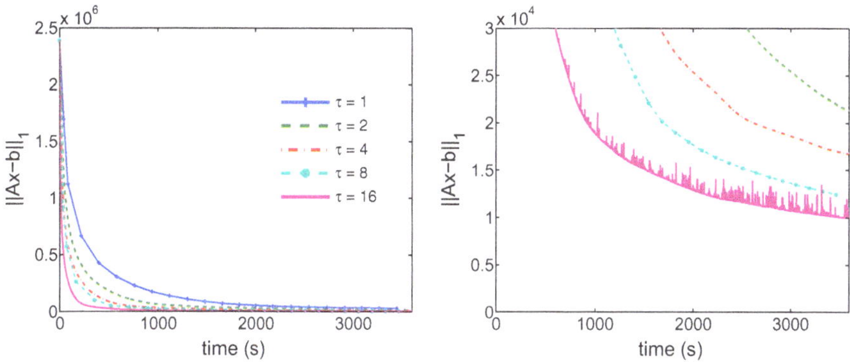

Fig. 2 Performance of SPCDM on the problem of minimizing $f(x) = \|Ax - b\|_1$ where A and b are given by the URL reputation dataset. We have run the method for $\tau \in \{1, 2, 4, 8, 16\}$ until the function value was decreased by a factor of 240. **Left**: decrease of the objective value in time. We can see that the parallelization speedup is proportional to the number of processors. **Right**: zoom on smaller objective values. We can see that the algorithm is not monotonic but monotonic on average

of $f(x) = \max_{1 \le j \le m} b_j(Ax)_j$ with $\mu = 1$. Hence, we can minimize f_1 by parallel coordinate descent with τ-nice sampling and β given by (54).

Convergence of AdaBoost is not a trivial result because the minimizing sequences may be unbounded. The proof relies on a decomposition of the optimization variables to an unbounded part and a bounded part [16, 36]. The original result gives iteration complexity $O(\frac{1}{\epsilon})$.

Parallel versions of AdaBoost have previously been studied. In our notation, Collins, Shapire and Singer [3] use $\tau = n$ and $\beta = \omega$. Palit and Reddy [24] use a generalized greedy sampling and take $\beta = \tau$ (number of processors). In the present work, we use randomized samplings and we can take $\beta \ll \min\{\omega, \tau\}$ with the τ-nice sampling. As discussed before, this value of β can be $O(\sqrt{n})$ times smaller than $\min\{\omega, \tau\}$, which leads to big gains in iteration complexity. For a detailed study of the properties of the SPCDM method applied to the AdaBoost problem, we refer to [6].

Experiment. In our last experiment, we demonstrate how SPCDM (which can be viewed as a random parallel version of AdaBoost) performs on the URL reputation dataset. Looking at Fig. 3, we see that parallelization leads to acceleration, and the time needed to decrease the loss to -1.85 is inversely proportional to the number of processors. Note that the additional effort done by increasing the number of processors from 4 to 8 is compensated by the increase of β from 1.2 to 2.0 (this is the little step in the zoom of Fig. 1). Even so, further acceleration takes place when one further increases the number of processors.

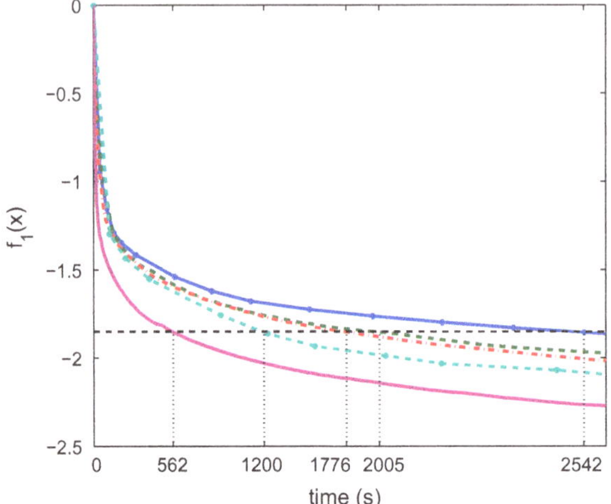

Fig. 3 Performance of the smoothed parallel coordinate descent method (SPCDM) with $\tau = 1, 2, 4, 8, 16$ processors, applied to the problem of minimizing the logarithm of the exponential loss (64), where $A \in \mathbb{R}^{m\times n}$ and $b \in \mathbb{R}^m$ are given by the URL reputation dataset; $m = 7{,}792{,}260$, $n = 3{,}231{,}961$ and $\omega = 414$. When $\tau = 16$ processors were used, the method needed 562s to obtain a solution of a given accuracy (depicted by the horizontal line). When $\tau = 8$ processors were used, the method needed 1200s, roughly double that time. Compared to a single processor, which needed 2542s, the setup with $\tau = 16$ was nearly 5 times faster. Hence, it is possible to observe nearly linear parallelization speedup, as our theory predicts. Same colors were used as in Fig. 2

Acknowledgements The work of both authors was supported by the EPSRC grant EP/I017127/1 (Mathematics for Vast Digital Resources). The work of P.R. was also supported by the Centre for Numerical Algorithms and Intelligent Software (funded by EPSRC grant EP/G036136/1 and the Scottish Funding Council).

References

1. Bradley, J.K., Kyrola, A., Bickson, D., Guestrin, C.: Parallel coordinate descent for L1-regularized loss minimization. In: 28th International Conference on Machine Learning (2011)
2. Bian, Y., Li, X., Liu, Y.: Parallel coordinate descent newton for large-scale L1-regularized minimization. arXiv:1306:4080v1 (2013)
3. Collins, M., Schapire, R.E., Singer, Y.: Logistic regression, adaboost and bregman distances. Mach. Learn. **48**(1–3), 253–285 (2002)
4. Devolder, O., Glineur, F., Nesterov, Y.: First-order methods of smooth convex optimization with inexact oracle. Math. Prog. 1–39 (2013)
5. Dang, C.D., Lan, G: Stochastic block mirror descent methods for nonsmooth and stochastic optimization. SIAM J. Opt. **25**(2), 856–881 (2015)
6. Fercoq, O.: Parallel coordinate descent for the AdaBoost problem. In: International Conference on Machine Learning and Applications—ICMLA '13 (2013)
7. Fercoq, O., Richtárik, P.: Accelerated, parallel, and proximal coordinate descent. SIAM J. Opt. **25**(4), 1997–2023 (2015)

8. Freund, Y., Schapire, R.E.: A decision-theoretic generalization of on-line learning and an application to boosting. In: Computational Learning Theory, pp. 23–37. Springer (1995)
9. Guyon, I., Gunn, S., Ben-Hur, A., Dror, G.: Result analysis of the NIPS 2003 feature selection challenge. Adv. Neural Inf. Process. Syst. **17**, 545–552 (2004)
10. Journée, M., Nesterov, Y., Richtárik, P., Sepulchre, R.: Generalized power method for sparse principal component analysis. J. Mach. Learn. Res. **11**, 517–553 (2010)
11. Lacoste-Julien, S., Jaggi, M., Schmidt, M., Pletcher, P.: Block-coordinate frank-wolfe optimization for structural svms. In: 30th International Conference on Machine Learning (2013)
12. Leventhal, D., Lewis, A.S.: Randomized methods for linear constraints: convergence rates and conditioning. Math. Op. Res. **35**(3), 641–654 (2010)
13. Liu, J., Wright, S.J., Ré, C., Bittorf, V., Sridhar, S.: An asynchronous parallel stochastic coordinate descent algorithm. J. Mach. Learn. Res. **16**(285–322), 1–5 (2015)
14. Liu, Z., Xiao, L.: On the complexity analysis of randomized block-coordinate descent methods. Math. Prog. **152**(1–2), 615–642 (2015)
15. Mukherjee, I., Canini, K., Frongillo, R., Singer, Y.: Parallel boosting with momentum. In: Lecture Notes in Computer Science, vol. 8188. Machine Learning and Knowledge Discovery in Databases, ECML (2013)
16. Mukherjee, I., Rudin, C., Schapire, R.E.: The rate of convergence of AdaBoost. J. Mach. Learn. Res. **14**(1), 2315–2347 (2013)
17. Ma, J., Saul, L.K., Savage, S., Voelker, G.M.: Identifying suspicious urls: an application of large-scale online learning. In: Proceedings of the 26th International Conference on Machine Learning, pp. 681–688. ACM (2009)
18. Necoara, I., Clipici, D.: Efficient parallel coordinate descent algorithm for convex optimization problems with separable constraints: application to distributed mpc. J. Process Control **23**, 243–253 (2013)
19. Nesterov, Y.: Smooth minimization of nonsmooth functions. Math. Prog. **103**, 127–152 (2005)
20. Nesterov, Y.: Efficiency of coordinate descent methods on huge-scale optimization problems. SIAM J. Opt. **22**(2), 341–362 (2012)
21. Nesterov, Y.: Subgradient methods for huge-scale optimization problems. Math. Prog. **146**(1), 275–297 (2014)
22. Necoara, I., Nesterov, Y., Glineur, F.: Efficiency of randomized coordinate descent methods on optimization problems with linearly coupled constraints. Politehnica Univ. of Bucharest, Technical report (2012)
23. Necoara, I., Patrascu, A.: A random coordinate descent algorithm for optimization problems with composite objective function and linear coupled constraints. Comput. Opt. Appl. **57**(2), 307–337 (2014)
24. Palit, I., Reddy, C.K.: Scalable and parallel boosting with MapReduce. IEEE Trans. Knowl. Data Eng. **24**(10), 1904–1916 (2012)
25. Richtárik, P., Takáč, M.: Efficiency of randomized coordinate descent methods on minimization problems with a composite objective function. In: 4th Workshop on Signal Processing with Adaptive Sparse Structured Representations, June 2011
26. Richtárik, P., Takáč, M.: Efficient serial and parallel coordinate descent methods for huge-scale truss topology design. In: Operations Research Proceedings, pp. 27–32. Springer (2012)
27. Richtárik, P., Takáč, M.: Iteration complexity of randomized block-coordinate descent methods for minimizing a composite function. Math. Prog. **144**(2), 1–38 (2014)
28. Richtárik, P., Takáč, M.: Parallel coordinate descent methods for big data optimization. Math. Prog. **156**(1), 433–484 (2016)
29. Richtárik, P., Takáč, M., Damla Ahipaşaoğlu, S.: Alternating maximization: unifying framework for 8 sparse PCA formulations and efficient parallel codes. arXiv:1212:4137 (2012)
30. Ruszczyński, A.: On convergence of an augmented Lagrangian decomposition method for sparse convex optimization. Math. Op. Res. **20**(3), 634–656 (1995)
31. Schapire, R.E., Freund, Y.: Boosting: Foundations and Algorithms. The MIT Press (2012)
32. Shalev-Shwartz, S., Tewari, A.: Stochastic methods for ℓ_1-regularized loss minimization. J. Mach. Learn. Res. **12**, 1865–1892 (2011)

33. Shalev-Shwartz, S., Zhang, T.: Accelerated mini-batch stochastic dual coordinate ascent. In: Advances in Neural Information Processing Systems, pp. 378–385 (2013)
34. Shalev-Shwartz, S., Zhang, T.: Stochastic dual coordinate ascent methods for regularized loss minimization. J. Mach. Learn. Res. **14**, 567–599 (2013)
35. Takáč, M., Bijral, A., Richtárik, P., Srebro, N.: Mini-batch primal and dual methods for SVMs. In: 30th International Conference on Machine Learning (2013)
36. Telgarsky, M.: A primal-dual convergence analysis of boosting. J. Mach. Learn. Res. **13**, 561–606 (2012)
37. Tao, Q., Kong, K., Chu, D., Wu, G.: Stochastic coordinate descent methods for regularized smooth and nonsmooth losses. In: Machine Learning and Knowledge Discovery in Databases, pp. 537–552 (2012)
38. Tappenden, R., Richtárik, P., Büke, B.: Separable approximations and decomposition methods for the augmented Lagrangian. Opt. Methods Softw. **30**(3), 643–668 (2015)
39. Tappenden, R., Richtárik, P., Gondzio, J.: Inexact coordinate descent: complexity and preconditioning. J. Opt. Theory Appl. 1–33 (2016)
40. Tappenden, R., Takáč, M., Richtárik, P.: On the complexity of parallel coordinate descent. Opt. Methods Softw. 1–24 (2017)

Olivier Fercoq recieved his Ph.D. from Ecole Polytechnique, France, in 2012. He then spent two years as a postdoctoral researcher at the University of Edinburgh (UK), where he met Peter Richtárik. He studied randomized coordinate descent methods for minimizing nonsmooth. In 2014, he chose to continue his research on optimization and randomized algorithms at Telecom ParisTech, in France, in the position of an assistant professor.

Peter Richtárik obtained his Ph.D. from Cornell University, USA, in 2007. Subsequently, he spent two years as a postdoctoral researcher at Universite catholique de Louvain, where he worked with Yurii Nesterov. Since 2009, he has been working at the University of Edinburgh, first as an assistant professor (lecturer), and later as an associate professor (reader). He is interested in randomized, parallel and distributed optimization, and the intersection of these fields with machine learning and linear algebra. The results presented in this paper came out of early collaboration between Fercoq and Richtárik, conducted in Edinburgh. While the performance of parallel coordinate descent methods was understood in the case of smooth (or composite) convex optimization, it was not clear what drives the performance of these methods for more complex nonsmooth problems. This paper gives the first results in this direction for functions of a max-type structure.

Second-Order Cone Optimization Formulations for Service System Design Problems with Congestion

Julio C. Góez and Miguel F. Anjos

Abstract We consider the service system design problem with congestion. This problem arises in a number of practical applications in the literature and is concerned with determining the location of service centers, the capacity level of each service center, and the assignment of customers to the facilities so that the customers demands are satisfied at total minimum cost. We present seven mixed-integer second-order cone optimization formulations for this problem, and compare their computational performances between them, and with the performance of other exact methods in the literature. Our results show that the conic formulations are competitive and may outperform the current leading exact methods. One advantage of the conic approach is the possibility of using off-the-shelf state-of-the-art solvers directly. More broadly, this study provides insights about different conic modeling approaches and the significant impact of the choice of approach on the efficiency of the resulting model.

Keywords Service systems · Congestion · Second-order cone optimization · Mixed-integer optimization

1 Introduction

We consider the service system design problem with congestion. In this problem, one has m different potential locations, and in each location there is a set of possible service capacities to satisfy incoming demand. We refer to those as service centers and their capacities are determined by the type of servers installed in each of them.

J. C. Góez (✉)
Department of Business and Management Science, NHH Norwegian School of Economics, Bergen, Norway
e-mail: julio.goez@nhh.no

M. F. Anjos
Canada Research Chair in Discrete Nonlinear Optimization in Engineering, GERAD & Polytechnique Montreal, Montreal, QC, Canada
e-mail: anjos@stanfordalumni.org

J. D. Pintér and T. Terlaky (eds.), *Modeling and Optimization: Theory and Applications*, Springer Proceedings in Mathematics & Statistics 279,
https://doi.org/10.1007/978-3-030-12119-8_5

We assumed that for all service centers, there are n different types of possible servers, and the service times for those servers are assumed to have exponential distributions. The cost of opening a service locations depends both on the location and the service capacity chosen. Additionally, there is a set of ℓ locations where the costumers' demand is originated, we refer to those as demand sources. We assume that every demand source may be served from any potential service center, and the cost of serving a demand source will depend on which service center is assigned to it. Also, we assume the orders generated at any demand source follows a Poisson process. The stochasticity both in the service facilities and the demand sources leads to a probability that the orders arriving to a service center will have to enter into a waiting queue, which will create congestion in the system. There is a cost per unit of time an order has to wait in the system before service. Hence, one has to decide which service facilities to open, with which type of servers, to ensure enough service capacity to satisfy the demand originated at the demand sources. The goal is to find a design for the system that is stable and minimizes the cost of opening the service centers plus the assignment cost of clients to service centers plus the average cost of the delay of orders in the system due to congestion. Stability in the system is obtained when the average waiting time in a queue for any order in any service center is finite. This problem arises in a number of practical applications in the literature. For example, in the facility location literature, it is known as the facility location problem with stochastic customer demand and immobile servers [8, 9, 30, 31]. Another example is the location of emergency service facilities such as medical clinics and preventive healthcare facilities [32]. It is also used to allocate optimally servers at different cloud locations to improve user request response times [21].

Amiri [2] formulated this problem using mixed-integer nonlinear optimization as follows. Consider a set of customers locations $\{1, \ldots, \ell\}$ indexed by i, a set of possible locations for service centers $\{1, \ldots, m\}$ indexed by j, and a set of type of servers $\{1, \ldots, n\}$ for the facilities indexed by k. We denote by $\lambda_i \geq 0$ the mean demand rate of customers at location i, and by $\mu_{jk} \geq 0$ the mean service rate of server of type k at service center j. Let d be the delay cost per unit of time, p_{jk} the cost of opening a service center at location j with server type k, and c_{ij} the cost of assigning customers at location i to a service center in location j. Additionally, let y_{jk} be a binary variable representing the decision for the location of the facilities, which takes the value of 1 if a service center is open at location j with capacity k, and zero otherwise. Also, let x_{ij} be a binary variable representing the assignment of clients to the facilities, which takes the value of 1 if the demand at location i is served by service center j, and zero otherwise. Hence, we obtain the following mixed-integer nonlinear optimization model:

$$\min \sum_{i=1}^{\ell}\sum_{j=1}^{m} c_{ij}x_{ij} + d\sum_{j=1}^{m} \frac{\sum_{i=1}^{\ell}\lambda_i x_{ij}}{\sum_{k=1}^{n}\mu_{jk}y_{jk} - \sum_{i=1}^{\ell}\lambda_i x_{ij}} + \sum_{j=1}^{m}\sum_{k=1}^{n} p_{jk}y_{jk} \tag{1}$$

$$\text{s.t.} \sum_{k=1}^{n}\mu_{jk}y_{jk} - \sum_{i=1}^{\ell}\lambda_i x_{ij} \geq 0 \quad j = 1, \ldots, m \tag{2}$$

$$\sum_{k=1}^{n} y_{jk} \leq 1 \quad j = 1, \ldots, m \tag{3}$$

$$\sum_{j=1}^{m} x_{ij} = 1 \quad i = 1, \ldots, \ell \tag{4}$$

$$x_{ij}, y_{jk} \in \{0, 1\} \quad i = 1, \ldots, \ell; \quad j = 1, \ldots, m; \quad k = 1, \ldots, n \tag{5}$$

For the queuing delay Amiri assumed each server service center as an independent $M/M/1$ queue, obtaining the second term in the objective function (1). In (2)–(5), we have three sets of constraints. The first set of constraints (2), ensures that the service rate at service center j is greater than or equal to the arrival rate of orders. These constraints ensure the stability of the queuing system at the service centers $\{1, \ldots, m\}$. Note that any feasible solution to this problem must satisfy the inequalities in (2) strictly to have stable queues at each service center. If that is not the case, one of the denominators may be zero in the second summation of the objective function, driving the design cost of the system to infinity. The second set of constraints (3) forces to select at most one of the available server types at each service center. The third set of constraints (4) ensures that each customer location is assigned to only one service center.

Note that if $d < 0$, then the model will reward the queuing delay in the system. In general, this goes against the purpose of service system design. Furthermore, if $d < 0$ then the objective function (1) will go to minus infinity, preferring unstable solutions. For these reasons, we assume throughout this paper that $d \geq 0$. An important drawback of formulation (1)–(5) is that if one of the facilities is not open, the set of constraints (2) will lead to an indeterminate $0/0$ term in the second summation of (1).

Different solution approaches have been proposed in the literature for this problem. Amiri [2] proposed a heuristic based on Lagrangian relaxation. Elhedhli proposed an equivalent mixed-integer linear optimization (MILO) formulation [12], and designed an exact algorithm based on outer linear approximations [13]. This problem has been also used as a testing problem for several techniques used to solve mixed-integer nonlinear problems, including disjunctive cuts [22], perspective reformulations [16], outer–inner approximation [18], and strong-branching inequalities [23].

In this paper, we explore mixed-integer second-order cone optimization (MISOCO) as a practical option to solve the service system design problem with congestion. Our contribution is a set of equivalent MISOCO formulations for the problem (1)–(5) that avoid the indeterminate problem present in formulation (1)–(5) when one of the facilities is not open. This approach seeks also to profit from the recent methodological developments, see, e.g., [3–6, 11, 15, 24–26], and from the power of off-the-shelf solvers capable of solving MISOCO problems [17, 20, 27]. Our computational results show that the MISOCO formulations are competitive and may outperform the leading exact methods in the literature. More broadly, our study also provides insights about different MISOCO modeling approaches and the significant impact of the choice of approach on the efficiency of the resulting model. Independently, about the same time, the work by Ahmadi-Javid and Hoseinpour [1] developed MISOCO formulations for these problems using the general model of a $M/G/1$ queue.

This paper is structured as follows. In Sect. 2, we present the previously proposed MILO formulation in [13] and the outer linear approximation approach in [13]. The new MISOCO formulations are derived in Sects. 3, and 4 reports the results of our computational testing to determine the most efficient formulation. Finally, we provide some concluding remarks in Sect. 5.

2 Review of Existing Exact Approaches

In this section, we summarize two approaches from the literature that are specialized for problem (1)–(5). Section 2.1 presents the MILO reformulation derived in [12], and Sect. 2.2 discusses the outer linear approximation proposed in [13].

2.1 MILO Approach

One way to deal with the nonlinearity in (1) is to use the result in [14, 28] that allows to linearize the product of a binary variable and a continuous variable. This is the same approach that is suggested in [12].

We start by introducing new variables s_j for the service centers via the inequalities

$$\frac{\sum_{i=1}^{\ell} \lambda_i x_{ij}}{\sum_{k=1}^{n} \mu_{jk} y_{jk} - \sum_{i=1}^{\ell} \lambda_i x_{ij}} \leq s_j, \quad s_j \geq 0, \quad j = 1, \ldots, m. \tag{6}$$

Using these inequalities, we can substitute the second summation in (1) by $d \sum_{j=1}^{m} s_j$ and add (6) to the set of constraints. Due to the nonnegativity of the s_j-s and d, plus the minimization direction, the inequalities in (6) will always be binding at optimality. Also, notice that

$$\frac{\sum_{i=1}^{\ell} \lambda_i x_{ij}}{\sum_{k=1}^{n} \mu_{jk} y_{jk} - \sum_{i=1}^{\ell} \lambda_i x_{ij}} \leq s_j \Longleftrightarrow \sum_{i=1}^{\ell} \lambda_i x_{ij} \leq s_j \left(\sum_{k=1}^{n} \mu_{jk} y_{jk} - \sum_{i=1}^{\ell} \lambda_i x_{ij} \right), \tag{7}$$

where the equivalence follows from (2).

Let $\bar{s}$ be an upper bound for the value of the s_j variables. Also let $w_{jk} = s_j y_{jk}$ and $z_{ij} = s_j x_{ij}$. Then, because x_{ij} and y_{jk} are binary, equation $w_{jk} = s_j y_{jk}$ can be modeled as follows [14, 28]:

$$w_{jk} \leq y_{jk}\bar{s}, \quad w_{jk} \leq s_j, \quad y_{jk}\bar{s} + s_j - w_{jk} \leq \bar{s}.$$

Similarly, equation $z_{ij} = s_j x_{ij}$ can be modeled as:

$$z_{ij} \leq x_{ij}\bar{s}, \quad z_{ij} \leq s_j, \quad x_{ij}\bar{s} + s_j - z_{ij} \leq \bar{s}.$$

This leads to the following MILO problem:

$$\min \sum_{i=1}^{\ell}\sum_{j=1}^{m} c_{ij}x_{ij} + d\sum_{j=1}^{m} s_j + \sum_{j=1}^{m}\sum_{k=1}^{n} p_{jk}y_{jk} \tag{8}$$

$$\text{s.t.} \sum_{k=1}^{n} \mu_{jk}w_{jk} - \sum_{i=1}^{\ell} \lambda_i z_{ij} \geq \sum_{i=1}^{\ell} \lambda_i x_{ij} \quad j = 1, \ldots, m \tag{9}$$

$$w_{jk} \leq y_{jk}\bar{s},\ w_{jk} \leq s_j,\ y_{jk}\bar{s} + s_j - w_{jk} \leq \bar{s},\ j = 1, \ldots, m,\ k = 1, \ldots, n \tag{10}$$

$$z_{ij} \leq x_{ij}\bar{s},\ z_{ij} \leq s_j,\ x_{ij}\bar{s} + s_j - z_{ij} \leq \bar{s},\ i = 1, \ldots, \ell,\ j = 1, \ldots, m \tag{11}$$

$$s_j, w_{jk}, z_{ij} \geq 0 \quad i = 1, \ldots, \ell,\ j = 1, \ldots, m,\ k = 1, \ldots, n \tag{12}$$

$$(3)-(5)$$

Notice that the stability of the queuing systems at each service center is ensured by the equivalence between (9)–(12) and (7). On the one hand, we that $w_{jk} = 0$ if either $y_{jk} = 0$ or $s_j = 0$, and also $z_{ij} = 0$ if either $x_{ij} = 0$ or $s_j = 0$. On the other hand, for any $y_{jk} = 1$ we obtain that $w_{jk} = s_j$, and for any $x_{ij} = 1$, we obtain that $z_{ij} = s_j$. Hence, at any optimal solution we have that

$$\sum_{k=1}^{n} \mu_{jk}w_{jk} - \sum_{i=1}^{\ell} \lambda_i z_{ij} = 0 \Leftrightarrow s_j\left(\sum_{k=1}^{n} \mu_{jk}y_{jk} - \sum_{i=1}^{\ell} \lambda_i x_{ij}\right) = 0.$$

Henceforth, from (9), we obtain that $\sum_{i=1}^{\ell} \lambda_i x_{ij} = 0$ for any optimal solution where $s_j = 0$, and given the minimization direction this ensures that $\sum_{k=1}^{n} \mu_{jk}y_{jk} = 0$. Now, if $\sum_{i=1}^{\ell} \lambda_i x_{ij} > 0$ at an optimal solution, then $s_j > 0$. In this last case, if $\sum_{k=1}^{n} \mu_{jk}y_{jk} - \sum_{i=1}^{\ell} \lambda_i x_{ij} = 0$ then $s_j \to \infty$, which will drive the design cost to infinity.

Finally, this reformulation has two main drawbacks. First, the number of constraints increases significantly with the MILO reformulation. Second, a good estimation of the upper bound $\bar{s}$ is needed.

2.2 Elhedhli's Exact Algorithm

The approach in [13] proposes an outer linear approximation to solve (1)–(5). Elhedhli's approach starts with a relaxation of the original nonlinear problem to a MILO problem. This relaxation is then iteratively refined using linear cuts to approximate the nonlinear constraints in the original problem. This refinement continues until the procedure reaches a difference between the upper and lower bounds below a given tolerance $\epsilon > 0$. The purpose of this section is to review the approach in [13].

The approach developed in [13] also introduces the variables s_i but defines them as follows:

$$s_j = \frac{\sum_{i=1}^{\ell} \lambda_i x_{ij}}{\sum_{k=1}^{n} \mu_{jk} y_{jk} - \sum_{i=1}^{\ell} \lambda_i x_{ij}}. \tag{13}$$

This allows to define the following linear relation between s_j and the x_{ij} and y_{jk} variables:

$$\sum_{i=1}^{\ell} \lambda_i x_{ij} = \sum_{k=1}^{n} \mu_{jk} z_{jk}, \quad z_{jk} = \frac{s_j}{1+s_j} y_{jk}.$$

Thus, using (13), we obtain the following equivalent reformulation:

$$\begin{aligned} \min \; & \sum_{i=1}^{\ell} \sum_{j=1}^{m} c_{ij} x_{ij} + d \sum_{i=j}^{m} s_j + \sum_{j=1}^{m} \sum_{k=1}^{n} p_{jk} y_{jk} \\ \text{s.t.} \; & \sum_{k=1}^{n} \mu_{jk} z_{jk} - \sum_{i=1}^{\ell} \lambda_i x_{ij} = 0 \quad j = 1, \ldots, m \\ & z_{jk} = \frac{s_j}{1+s_j} y_{jk} \quad j = 1, \ldots, m, \quad k = 1, \ldots, n \\ & s_j \geq 0, \, j = 1, \ldots, m \\ & (3)-(5) \end{aligned} \tag{14}$$

An important feature of this reformulation is that it guarantees the stability of the queuing systems at each location. To see this, notice that since $s_j \geq 0$ the second set of constraints in (14) ensures that $z_{jk} < 1$ for $j = 1, \ldots, m$ and $k = 1, \ldots, n$. Then, the first set of constraints in (14) will force the rate of clients $\sum_{i=1}^{\ell} \lambda_i x_{ij}$ arriving to a service center j to be a fraction of the assigned service rate.

The nonlinearity of the problem is now found in the second constraint of (14). The basis for the approach in [13] is to exploit the fact that the function $f(s_j) = \frac{s_j}{1+s_j}$ is concave. Elhedhli's approach uses first-order approximations of the function $f(s_j)$ to reformulate the problem. In particular, using the set $\mathcal{H}$ of all $s_j^h \geq 0$ where $h \in \mathcal{H}$, then we have that:

$$\frac{s_j}{1+s_j} \leq \frac{1}{(1+s_j^h)^2} s_j + \left(\frac{s_j^h}{1+s_j^h} \right), \quad \forall h \in \mathcal{H}.$$

Using this result, we obtain the following equivalent reformulation of the problem [13]:

$$\begin{aligned}
\min \ & \sum_{i=1}^{\ell} \sum_{j=1}^{m} c_{ij} x_{ij} + d \sum_{j=1}^{m} s_j + \sum_{j=1}^{m} \sum_{k=1}^{n} p_{jk} y_{jk} \\
\text{s.t.} \ & \sum_{k=1}^{n} \mu_{jk} z_{jk} - \sum_{i=1}^{\ell} \lambda_i x_{ij} = 0 \quad j = 1, \ldots, m, \\
& z_{jk} \leq y_{jk} \quad j = 1, \ldots, m; \quad k = 1, \ldots, n, \\
& z_{jk} - \frac{1}{(1+s_j^h)^2} s_j \leq \left(\frac{s_j^h}{1+s_j^h} \right)^2 \quad j = 1, \ldots, m; \; k = 1, \ldots, n; \; h \in \mathcal{H}, \\
& s_j \geq 0, \quad j = 1, \ldots, m, \\
& z_{jk} \geq 0, \quad j = 1, \ldots, m; \quad k = 1, \ldots, n. \\
& (3){-}(5).
\end{aligned} \tag{15}$$

Note that $|\mathcal{H}| = \infty$ in (15). For that reason, the approach in [13] proposes to start with a relaxation of (15) in which a finite subset $\hat{\mathcal{H}} \subseteq \mathcal{H}$ is used, and new s_j^h are added to $\hat{\mathcal{H}}$ iteratively such that the relaxation is refined at each iteration of the procedure. This procedure is summarized in Algorithm 1.

An important remark about this procedure is necessary. Note that the solution of a relaxation of the problem (15) can be such that

$$\sum_{k=1}^{n} \mu_{\hat{j}k} y_{\hat{j}k}^* - \sum_{i=1}^{\ell} \lambda_i x_{i\hat{j}}^* = 0 \text{ for some } \hat{j}.$$

This makes UB tend to infinity in step 7 of Algorithm 1, which implies that the objective function of the original problem for the solution (x^*, y^*) is $+\infty$. In other words, the solution (x^*, y^*) results in at least one unstable queue system at one of the facilities. This may happen because the reformulation (15) guarantees stability only when the infinite set of $s_j^h \in \mathcal{H}$ values is considered for $j = 1, \ldots, m$.

Consider for example the case illustrated in Fig. 1, with four customer locations points D_1, D_2, D_3, and D_4, and two possible server locations, each with the option of having a service rate of 10 or 20. We assume that the connection from any demand point to any server has a cost of \$15, the installation of a server with a service rate of 10 costs \$100, and the installation of a server with a service rate of 20 has a cost of \$500. Finally, the cost associated with congestion is one dollar for each unit of time spent in the system. In this case, we may take advantage of the size of the problem to

Algorithm 1 Elhedhli's approach

1: $UB \leftarrow +\infty$
2: $LB \leftarrow -\infty$
3: $\mathcal{H} \leftarrow$ well-placed initial set of cuts
4: **while** $\frac{UB-LB}{UB} > \epsilon$ **do**
5: Solve problem (15) and let (x^*, y^*, z^*, s^*) be the optimal solution
6:
$$LB \leftarrow \sum_{i=1}^{\ell}\sum_{j=1}^{m} c_{ij}x_{ij}^* + d\sum_{j=1}^{m} s_j^* + \sum_{j=1}^{m}\sum_{k=1}^{n} p_{jk}y_{jk}^*$$
7:
$$UB \leftarrow \sum_{i=1}^{\ell}\sum_{j=1}^{m} c_{ij}x_{ij}^* + d\sum_{j=1}^{m} \frac{\sum_{i=1}^{\ell}\lambda_i x_{ij}^*}{\sum_{k=1}^{n}\mu_{jk}y_{jk}^* - \sum_{i=1}^{\ell}\lambda_i x_{ij}^*} + \sum_{j=1}^{m}\sum_{k=1}^{n} p_{jk}y_{jk}^*$$
8: **if** $\frac{UB-LB}{UB} > \epsilon$ **then**
9: **for** $j \leftarrow 1, \ldots, m$ **do**
10: $\mathcal{H} \leftarrow \mathcal{H} \cup \left\{ \frac{\sum_{i=1}^{\ell}\lambda_i x_{ij}^*}{\sum_{k=1}^{n}\mu_{jk}y_{jk}^* - \sum_{i=1}^{\ell}\lambda_i x_{ij}^*} \right\}$
11: **end for**
12: **end if**
13: **end while**

enumerate all the stable solutions for the problem in Fig. 1. Note that for any stable solution, the servers with service rate 10 can satisfy the demand of at most one of the demand sources. Also note that the servers with service rate 20 can satisfy the demand of at most three demand sources. Hence, the list of possible configurations for the servers in this problem for a solution to be stable are (μ_{11}, μ_{22}), (μ_{12}, μ_{21}), (μ_{12}, μ_{22}), where the first component is the capacity assigned to server 1 and the second is the capacity assigned to server 2.

With those configurations, we can compute all the possible values for the s_j that can be obtained running Algorithm 1. That characterizes all the possible cuts generated with Algorithm 1 because unstable solutions make step 10 fail. If we use those cuts as the initial well-place cuts, then Algorithm 1 can be initialized with the following MILO relaxation:

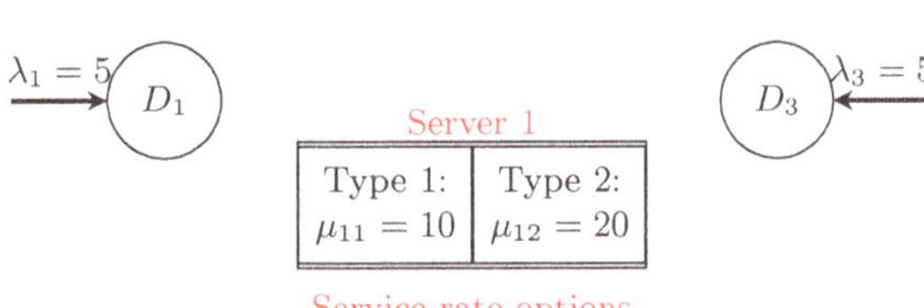

Fig. 1 Example of a service system design problem where step 10 of Algorithm 1 may fail

$$\min 15 \sum_{i=1}^{4}\sum_{j=1}^{2} x_{ij} + \sum_{j=1}^{2} s_j + 100 \sum_{j=1}^{2} y_{j1} + 500 \sum_{j=1}^{2} y_{j2} \tag{16}$$

$$\text{s.t. } 5 \sum_{i=1}^{4} x_{ij} - 10 z_{j1} - 20 z_{j2} = 0, \quad j = 1, 2 \tag{17}$$

$$\sum_{j=1}^{2} x_{ij} = 1, \quad i = 1, \ldots, 4 \tag{18}$$

$$\sum_{k=1}^{2} y_{jk} \leq 1, \quad j = 1, 2 \tag{19}$$

$$z_{jk} - y_{jk} \leq 0, \quad j = 1, 2, \quad k = 1, 2 \tag{20}$$

$$z_{jk} - \frac{1}{4} s_j \leq \frac{1}{4}, \quad j = 1, 2, \quad k = 1, 2 \tag{21}$$

$$z_{jk} - \frac{1}{16} s_j \leq \frac{9}{16}, \quad j = 1, 2, \quad k = 1, 2 \tag{22}$$

$$z_{jk} - \frac{9}{16} s_j \leq \frac{1}{16}, \quad j = 1, 2, \quad k = 1, 2 \tag{23}$$

$$x_{ij}, y_{jk} \in \{0, 1\}, \quad i = 1, \ldots, 4, \quad j = 1, 2, \quad k = 1, 2 \tag{24}$$

$$s_j, z_{jk} \geq 0, \quad j = 1, 2, \quad k = 1, 2. \tag{25}$$

Analyzing the structure of Problem (16)–(25), the first observation is that the first term in the objective function can be treated as a constant. The second observation is that the difference in cost between servers with capacity 10 and servers with capacity 20 is 400. This tells us that to use a server of capacity 20, one need savings of 400 or more. The third observation is that those savings must come from lowering the time the clients spend in the system, which is in this case given by the s_j. Given this structure, it follows that there are 12 optimal solutions to problem (16)–(25) with optimal objective value equal to 290. The common structure for those solutions is that all of them will use the two servers of capacity 10. Example of an optimal solution is:

$$x_{11} = x_{12} = x_{23} = x_{24} = 1, y_{11} = z_{11} = y_{21} = z_{21} = 1, s_1 = s_2 = 7,$$

where the rest of the variables would be zero an omitted for the sake of clarity. Note that these solutions are suboptimal for the original problem. In this situation, it is not possible to compute new values for s_j^h, which makes impossible to generate new linear cuts. Hence, the algorithm will finish with a suboptimal solution.

In summary, Algorithm 1 ignores that it is possible that no new s_j^h may be computed. We note that in practice, with a careful approach for the definition of the initial cuts, this problem may be in general avoided. For example, in [12, 29], the initial set of cuts is chosen so that the approximation achieved with the tangents is arbitrarily precise. However, that still does not guarantee the elimination of unstable solutions.

3 New MISOCO Formulations

Before deriving our MISOCO formulations, we need to reformulate the objective function of Problem (1)–(5), which can lead to the indeterminate form $\frac{0}{0}$ in the second term when a service center is not deployed. To avoid this situation, we linearize that objective function as follows. Consider the nonlinear term in the objective function (1)

$$d \sum_{j=1}^{m} \frac{\sum_{i=1}^{\ell} \lambda_i x_{ij}}{\sum_{k=1}^{n} \mu_{jk} y_{jk} - \sum_{i=1}^{\ell} \lambda_i x_{ij}}.$$

Using again the inequalities (6), we obtain the following formulation:

$$\min \sum_{i=1}^{\ell} \sum_{j=1}^{m} c_{ij} x_{ij} + d \sum_{j=1}^{m} \sum_{j=1}^{m} s_j + \sum_{j=1}^{m} \sum_{k=1}^{n} p_{jk} y_{jk} \tag{26}$$

$$\text{s.t. } s_j \geq \begin{cases} \frac{\sum_{i=1}^{\ell} \lambda_i x_{ij}}{\sum_{k=1}^{n} \mu_{jk} y_{jk} - \sum_{i=1}^{\ell} \lambda_i x_{ij}}, & \text{if } \sum_{i=1}^{\ell} \lambda_i x_{ij} > 0 \\ 0, & \text{otherwise.} \end{cases} \quad j = 1, \ldots, m \tag{27}$$

$$(2)-(5)$$

In the case when all the facilities considered in the design will be forced to be open, then formulation (26)–(27), (2)–(5) is equivalent to (1)–(5). To see this, we need to show that (26)–(27), (2)–(5) does not ignore the effect that the cost d has in the objective function (1) for unstable solutions. For that, we need to focus on the case when $\sum_{k=1}^{n} \mu_{jk} y_{jk} - \sum_{i=1}^{\ell} \lambda_i x_{ij} = 0$ for some server j. There are two different cases in which this may happen. First, consider the case when $\sum_{j=1}^{n} \lambda_i x_{ij} = 0$. Because (26) is minimizing, we obtain that $s_j = 0$ and $\sum_{k=1}^{n} \mu_{jk} y_{jk} = 0$, which is a stable solution. Second, consider the case when $\sum_{j=1}^{n} \lambda_i x_{ij} = \sum_{k=1}^{n} \mu_{jk} y_{jk} > 0$. Then to satisfy (27), we need that $s_j \to +\infty$, which makes (26) go to $+\infty$. This shows that the effect of the cost d in the original formulation for unstable solutions is still accounted for in (26)–(27), (2)–(5). This equivalence highlights the fact that the initial formulation can't handle the decision of not opening a specific service center. That is the main practical contribution of (26)–(27), (2)–(5).

We may now use this reformulation to derive several equivalent MISOCO formulations. We start in Sect. 3.4 with what we consider straightforward MISOCO formulations where we use the binary constraint on the x vector. Then, we show in Sect. 3.4 that it is also possible to use the y vector to obtain alternative MISOCO formulations. These two approaches use Eq. (27) to obtain a MISOCO reformulation. However, the choice of which binary variable to use will determine the dimension and the number of SOC in the MISOCO obtained. Also, for the two approaches we present a reformulation option that fixes the dimensions of the SOC to 3 at the cost of increasing the number of SOCs in the formulation. Finally, in Sect. 3.3, we show how alternative formulations may be obtained by introducing the traffic intensity of an $M/M/1$ queuing system as a new variable in the formulation. With this approach, we can express the nonlinearity in the objective function in terms of the traffic intensity. This allows the derivation of MISOCO reformulations that avoid the use of the stability constraint in the derivation of the SOC.

3.1 MISOCO Models Based on the x Variables

Our first MISOCO formulations arise from using the binary nature of x to deal with the nonlinearity in constraint (27).

MISOCO Formulation 1

In this first approach, we manage the nonlinearity in (27) as follows. First, from (7) and the binary constraint over the x vector, we have that

$$\frac{\sum_{i=1}^{\ell} \lambda_i x_{ij}}{\sum_{k=1}^{n} \mu_{jk} y_{jk} - \sum_{i=1}^{\ell} \lambda_i x_{ij}} \leq s_j \quad \Longleftrightarrow \quad \sum_{i=1}^{\ell} \lambda_i x_{ij}^2 \leq s_j \left(\sum_{k=1}^{n} \mu_{jk} y_{jk} - \sum_{i=1}^{\ell} \lambda_i x_{ij} \right). \tag{28}$$

Notice that thanks to the minimization direction in (26), the right-hand side inequality in (28) captures all the properties of (27). In other words, if $\sum_{i=1}^{\ell} \lambda_i x_{ij} = 0$ then $s_j = 0$, and if $\sum_{i=1}^{\ell} \lambda_i x_{ij} > 0$ then $s_j \to \infty$ when $\sum_{k=1}^{n} \mu_{jk} y_{jk} - \sum_{i=1}^{\ell} \lambda_i x_{ij} \to 0$.

Next, we introduce a variable $t_j \geq 0$ to obtain the following reformulation of the inequality on the right-hand side of (28):

$$\sum_{i=1}^{\ell} \lambda_i x_{ij}^2 \leq s_j t_j \tag{29}$$

$$\left(\sum_{k=1}^{n} \mu_{ij} y_{jk} - \sum_{i=1}^{\ell} \lambda_i x_{ij} \right) \geq t_j. \tag{30}$$

Notice that the inequality on the right of (28) and (29)–(30) are not equivalent in general. However, in this context, their equivalence comes from the minimizing direction of the problem and the assumption that $d \geq 0$; this forces (30) to be always binding. Also notice that (30) dominates (2) and that (29) is a rotated second-order cone (SOC). Hence, we obtain our first MISOCO reformulation for problem (26)–(27), (2)–(5):

$$
\begin{aligned}
\min \quad & \sum_{i=1}^{\ell}\sum_{j=1}^{m} c_{ij}x_{ij} + d\sum_{j=1}^{m} s_j + \sum_{j=1}^{m}\sum_{k=1}^{n} p_{jk}y_{jk} \\
\text{s.t.} \quad & \sum_{k=1}^{n} \mu_{jk}y_{jk} - \sum_{i=1}^{\ell} \lambda_i x_{ij} \geq t_j \quad j = 1, \ldots, m \\
& \sum_{i=1}^{\ell} \lambda_i x_{ij}^2 \leq s_j t_j \quad j = 1, \ldots, m \\
& t_j, s_j \geq 0 \quad j = 1, \ldots, m \\
& (3)-(5).
\end{aligned}
\tag{MISOCO 1}
$$

From this reformulation process, we obtain the following result.

Lemma 1 *The formulation* (MISOCO 1) *is equivalent to the problem formulation* (26)–(27), (2)–(5).

The dimensions of the problem affect formulation (MISOCO 1) as follows:

- The number of SOCs depends on the number of service centers m, which is usually significantly smaller than the number of demand locations.
- The dimension of the cones depends on the number of demand locations ℓ.

MISOCO Formulation 2

Here, we manage the nonlinear constraint (27) in a slightly different way. Because

$$
\frac{\sum_{i=1}^{\ell} \lambda_i x_{ij}}{\sum_{k=1}^{n} \mu_{jk}y_{jk} - \sum_{i=1}^{\ell} \lambda_i x_{ij}} = \sum_{p=1}^{\ell} \frac{\lambda_p x_{pj}}{\sum_{k=1}^{n} \mu_{jk}y_{jk} - \sum_{i=1}^{\ell} \lambda_i x_{ij}},
$$

we can reformulate the left-hand side inequality of (28) as:

$$
\sum_{i=1}^{\ell} u_{ij} \leq s_j, \quad \frac{\lambda_i x_{ij}}{\sum_{k=1}^{n} \mu_{jk}y_{jk} - \sum_{i=1}^{\ell} \lambda_i x_{ij}} \leq u_{ij}. \tag{31}
$$

Note that the minimization will force the second inequality in (31) to be binding. Furthermore, using the binary constraint over x and (2), we obtain

$$
\frac{\lambda_i x_{ij}}{\sum_{k=1}^{n} \mu_{jk}y_{jk} - \sum_{i=1}^{\ell} \lambda_i x_{ij}} \leq u_{ij} \quad \Longleftrightarrow \quad \lambda_i x_{ij}^2 \leq u_{ij}\left(\sum_{k=1}^{n} \mu_{jk}y_{jk} - \sum_{i=1}^{\ell} \lambda_i x_{ij}\right). \tag{32}
$$

Again, the inequality on the right in (32) captures all the properties of (27). Here, we also introduce auxiliary variables t_j to reformulate the rotated SOC (32), and hence obtain our second MISOCO reformulation for problem (26)–(27), (2)–(5):

$$\begin{aligned}
\min \; & \sum_{i=1}^{\ell}\sum_{j=1}^{m} c_{ij}x_{ij} + d\sum_{j=1}^{m} s_j + \sum_{j=1}^{m}\sum_{k=1}^{n} p_{jk}y_{jk} \\
\text{s.t.} \; & \sum_{k=1}^{n} \mu_{jk}y_{jk} - \sum_{i=1}^{\ell} \lambda_i x_{ij} \geq t_j \quad j = 1, \ldots, m \\
& \sum_{i=1}^{\ell} u_{ij} \leq s_j \quad j = 1, \ldots, m \\
& \lambda_i x_{ij}^2 \leq t_j u_{ij} \quad i = 1, \ldots, \ell; \quad j = 1, \ldots, m \\
& t_j, u_{ij} \geq 0 \quad i = 1, \ldots, \ell; \quad j = 1, \ldots, m \\
& (3)-(5).
\end{aligned} \qquad \text{(MISOCO 2)}$$

From this reformulation process, we obtain the following result.

Lemma 2 *The formulation* (MISOCO 2) *is equivalent to the problem formulation* (26)–(27), (2)–(5).

The dimensions of the problem affect formulation (MISOCO 2) as follows:

- The number of SOCs depends on the number of demand sources ℓ and service locations m. Specifically, for each additional demand source, we need to add m new rotated SOCs for a total of ℓm additional cones.
- On the other hand, the dimension of the cones is fixed to 3.

3.2 MISOCO Models Based on the y Variables

Our next group of MISOCO models shows how we can use the binary constraint over y to deal with the nonlinearity in constraint (27).

MISOCO Formulation 3

For this model, we start by adding the term $\sum_{k=1}^{n} \mu_{jk}y_{jk} - \sum_{i=1}^{\ell} \lambda_i x_{ij}$ to both sides of the inequality on the right-hand side of (7) to obtain

$$\sum_{k=1}^{n} \mu_{jk}y_{jk} \leq (1 + s_j)\left(\sum_{k=1}^{n} \mu_{jk}y_{jk} - \sum_{i=1}^{\ell} \lambda_i x_{ij}\right). \tag{33}$$

Using the binary constraint over y and (2), the inequality (33) can be reformulated as a rotated SOC as follows:

$$\sum_{k=1}^{n} \mu_{jk} y_{jk}^2 \leq (1 + s_j) \left(\sum_{k=1}^{n} \mu_{jk} y_{jk} - \sum_{i=1}^{\ell} \lambda_i x_{ij} \right). \tag{34}$$

Similarly to what we did in section "MISOCO Formulation 1", we may introduce an auxiliary variable t_j to reformulate the rotated SOC (34). Hence, we obtain our third MISOCO reformulation for problem (26)–(27), (2)–(5):

$$\begin{aligned} \min \ & \sum_{i=1}^{\ell} \sum_{j=1}^{m} c_{ij} x_{ij} + d \sum_{j=1}^{m} s_j + \sum_{j=1}^{m} \sum_{k=1}^{n} p_{jk} y_{jk} \\ \text{s.t.} \ & \sum_{k=1}^{n} \mu_{ik} y_{ik} - \sum_{i=1}^{\ell} \lambda_i x_{ij} \geq t_j \quad j = 1, \ldots, m \\ & \sum_{k=1}^{n} \mu_{jk} y_{jk}^2 \leq (1 + s_j) t_j \quad j = 1, \ldots, m \\ & t_j \geq 0 \quad j = 1, \ldots, m \\ & (3)-(5). \end{aligned} \tag{MISOCO 3}$$

From this reformulation process, the following result follows.

Lemma 3 *The formulation* (MISOCO 3) *is equivalent to the problem formulation* (26)–(27), (2)–(5).

The dimensions of the problem affect formulation (MISOCO 3) as follows:

- The number of SOCs depends on the number of service centers m.
- The dimensions of the cones depend on the number of capacity levels n.

Furthermore, we note that the rotated cones have smaller dimension than for (MISOCO 1) because this formulation works on the y variables instead of the x variables (in a way similar to what was done in section "MISOCO Formulation 1").

MISOCO Formulation 4

For this model, we use the constraint (3) and the rotated SOC constraint in (MISOCO 3) to derive a new MISOCO formulation. Recall the constraint

$$\sum_{k=1}^{n} y_{jk} \leq 1 \quad j = 1, \ldots, m, \tag{35}$$

which requires that for any feasible solution at most one server level is deployed. Hence, the rotated SOC constraint $\sum_{k=1}^{n} \mu_{jk} y_{jk}^2 \leq (1 + s_j) t_j$ in (MISOCO 3) is equivalent to the following set of constraints

$$\mu_{jk}y_{jk}^2 \le (1+s_j)t_j \quad j=1,\dots,m; \quad k=1,\dots,n. \tag{36}$$

Note that if $\mu_{jk}=0$, then capacity k for server j is zero and that server level can simply be ignored in the formulation.

We thus obtain our fourth MISOCO reformulation for problem (26)–(27), (2)–(5):

$$\begin{aligned}
\min\ & \sum_{i=1}^{\ell}\sum_{j=1}^{m} c_{ij}x_{ij} + d\sum_{j=1}^{m} s_j + \sum_{j=1}^{m}\sum_{k=1}^{n} p_{jk}y_{jk} && \text{(MISOCO 4)}\\
\text{s.t. } & \sum_{k=1}^{n}\mu_{jk}y_{jk} - \sum_{i=1}^{\ell}\lambda_i x_{ij} \ge t_j \quad j=1,\dots,m\\
& \mu_{jk}y_{jk}^2 \le (1+s_j)t_j \quad j=1,\dots,m; \quad k=1,\dots,n\\
& t_j \ge 0 \quad j=1,\dots,m\\
& (3)-(5).
\end{aligned}$$

From this reformulation process, the following result follows.

Lemma 4 *The formulation* (MISOCO 4) *is equivalent to the problem formulation* (26)–(27), (2)–(5).

The dimensions of the problem affect formulation (MISOCO 4) as follows:

- The number of SOCs depends on the number of service centers m and levels n. Specifically, with each additional level, we need to include m new rotated cones.
- The advantage that the dimension of these cones is fixed to 3.

MISOCO Formulation 5

For our fifth formulation, we want to exploit further the binary constraint on the variables y and constraint (3). We first define the variables:

$$u_j = \sum_{k=1}^{n}\sqrt{\mu_{jk}}y_{jk}, \quad j=1,\dots,m. \tag{37}$$

and use them to reformulate (33) for each $j=1,\dots,m$ as follows:

$$u_j^2 \le (1+s_j)t_j. \tag{38}$$

Hence, we obtain our fifth MISOCO reformulation for problem (26)–(27), (2)–(5):

$$\begin{aligned}\min\ & \sum_{i=1}^{\ell}\sum_{j=1}^{m} c_{ij}x_{ij} + d\sum_{j=1}^{m} s_j + \sum_{j=1}^{m}\sum_{k=1}^{n} p_{jk}y_{jk} \\ \text{s.t. } & \sum_{k=1}^{n}\mu_{jk}y_{jk} - \sum_{i=1}^{\ell}\lambda_i x_{ij} \ge t_j \quad j=1,\ldots,m \\ & u_j^2 \le (1+s_j)t_j \quad j=1,\ldots,m \\ & u_j = \sum_{k=1}^{n}\sqrt{\mu_{jk}}y_{jk},\ j=1,\ldots,m \\ & t_j, u_j \ge 0 \quad j=1,\ldots,m \\ & (3)-(5).\end{aligned} \tag{MISOCO 5}$$

From this reformulation process, the following result follows.

Lemma 5 *The formulation* (MISOCO 5) *is equivalent to the problem formulation* (26)–(27), (2)–(5).

The dimensions of the problem affect formulation (MISOCO 5) as follows:

- The number of SOCs depends on the number of service centers m.
- The dimension of the cones is fixed to 3.

Notice that this formulation does not require the direct use of the binary constraint over y to obtain the rotated SOC.

3.3 MISOCO Models Based on the Traffic Intensity

For our two final MISOCO formulations, we use the traffic intensity of an $M/M/1$ queuing system defined (in our case) as:

$$\rho_j = \frac{\sum_{i=1}^{\ell}\lambda_i x_{ij}}{\sum_{k=1}^{n}\mu_{jk}y_{jk}}, \quad j=1,\ldots,m. \tag{39}$$

Using (39), we can reformulate (26)–(27), (2)–(5) as follows:

$$\min \sum_{i=1}^{\ell}\sum_{j=1}^{m} c_{ij}x_{ij} + d\sum_{j=1}^{m}\sum_{j=1}^{m} s_j + \sum_{j=1}^{m}\sum_{k=1}^{n} p_{jk}y_{jk} \tag{40}$$

$$\text{s.t. } \frac{\rho_j}{1-\rho_j} \le s_j \quad j=1,\ldots,m \tag{41}$$

$$\rho_j \ge \begin{cases}\frac{\sum_{i=1}^{\ell}\lambda_i x_{ij}}{\sum_{k=1}^{n}\mu_{jk}y_{jk}}, & \text{if } \sum_{i=1}^{\ell}\lambda_i x_{ij} > 0 \\ 0, & \text{otherwise.}\end{cases} \quad j=1,\ldots,m \tag{42}$$

$$0 \le \rho_j \le 1, s_j \ge 0 \quad j = 1, \ldots, m \tag{43}$$
$$(2)-(5).$$

The inequality in (42) follows from the minimization direction. Specifically, note that for some $\hat{\rho} < \bar{\rho}$ one has that

$$\frac{\hat{\rho}}{1-\hat{\rho}} < \frac{\bar{\rho}}{1-\bar{\rho}}.$$

Note also that here the possibility of an indeterminate form is handled with (42) and not expression $\frac{0}{0}$ will be encountered.

Now, we need to manage the nonlinearities in (41) and (42). For (41) we have that:

$$\frac{\rho_j}{1-\rho_j} \le s_j \Leftrightarrow \rho_j \le s_j\left(1-\rho_j\right) \Leftrightarrow 1 \le \left(1+s_j\right)\left(1-\rho_j\right),$$

where the last inequality is SOC-representable [7, Chap. 3], since $1-\rho_j \ge 0$ and $1+s_j \ge 0$.

For (42), we have that

$$\frac{\sum_{i=1}^{\ell} \lambda_i x_{ij}}{\sum_{k=1}^{n} \mu_{jk} y_{jk}} \le \rho_j \Leftrightarrow \sum_{i=1}^{\ell} \lambda_i x_{ij} \le \rho_j \sum_{k=1}^{n} \mu_{jk} y_{jk}. \tag{44}$$

Notice that in this formulation (44) is equivalent to (42). Note also that (44) dominates (2), which allows us to greatly simplify the problem formulation as follows:

$$\min \sum_{i=1}^{\ell}\sum_{j=1}^{m} c_{ij}x_{ij} + d\sum_{j=1}^{m} s_j + \sum_{j=1}^{m}\sum_{k=1}^{n} p_{jk}y_{jk} \tag{45}$$

$$\text{s.t.} \sum_{i=1}^{\ell} \lambda_i x_{ij} \le \rho_j \sum_{k=1}^{n} \mu_{jk} y_{jk} \quad j = 1, \ldots, m \tag{46}$$

$$1 \le \left(1+s_j\right)\left(1-\rho_j\right) \quad j = 1, \ldots, m \tag{47}$$

$$0 \le \rho_j \le 1, s_j \ge 0 \quad j = 1, \ldots, m \tag{48}$$

$$(3)-(5).$$

From this reformulation process, we obtain the following result.

Lemma 6 *The formulation* (45)–(48), *is equivalent to the problem formulation* (26)–(27), (2)–(5).

To obtain MISOCO reformulations using this path, it remains to deal with the nonlinearity in (46). We present two alternatives in the following sections.

MISOCO Formulation 6

The first alternative is to use the binary constraint on the x variables to obtain the following MISOCO reformulation:

$$
\begin{aligned}
\min \ & \sum_{i=1}^{\ell}\sum_{j=1}^{m} c_{ij}x_{ij} + d\sum_{j=1}^{m} s_j + \sum_{j=1}^{m}\sum_{k=1}^{n} p_{jk}y_{jk} \\
\text{s.t.} \ & \sum_{i=1}^{\ell} \lambda_i x_{ij}^2 \le \rho_j \sum_{k=1}^{n} \mu_{jk} y_{jk} \quad j = 1,\dots,m \\
& (3)\text{–}(5),\ (47)\text{–}(48).
\end{aligned}
\qquad \text{(MISOCO 6)}
$$

The dimensions of the problem affect formulation (MISOCO 6) as follows:

- The number of SOCs depends on the number of service centers m. Specifically, this formulation requires $2m$ rotated SOCs.
- The m rotated SOCs in (47) have fixed dimension 3, while the dimension of the rotated SOCs in the first constraint of (MISOCO 6) depends on the number of demand locations ℓ.

MISOCO Formulation 7

The second alternative is to consider the constraints

$$\sum_{k=1}^{n} y_{jk} \le 1, \quad 0 \le \rho_j \le 1,$$

and to introduce new variables $z_{jk} = \rho_j y_{jk}$. Using these for each j makes it possible to reformulate (46) as:

$$
\begin{aligned}
\min \ & \sum_{i=1}^{\ell}\sum_{j=1}^{m} c_{ij}x_{ij} + d\sum_{j=1}^{m} s_j + \sum_{j=1}^{m}\sum_{k=1}^{n} p_{jk}y_{jk} \\
\text{s.t.} \ & \sum_{i=1}^{\ell} \lambda_i x_{ij} \le \sum_{k=1}^{n} \mu_{jk} z_{jk} \quad j = 1,\dots,m \\
& \sum_{k=1}^{n} z_{jk} \le \rho_j \quad j = 1,\dots,m \\
& z_{jk} \le y_{jk} \quad j = 1,\dots,m; \quad k = 1,\dots,n \\
& (3)\text{–}(5),\ (47)\text{–}(48).
\end{aligned}
\qquad \text{(MISOCO 7)}
$$

Note that the minimization will always make the second set of constraints in (MISOCO 7) bounding.

The dimensions of the problem affect formulation (MISOCO 7) as follows:

- The number of rotated SOCs in this formulation equals the number of service centers m.
- The m rotated SOCs in (47) have fixed dimension 3.

Note that (MISOCO 7) is closely related to formulation (15). What differentiates the two formulations is the way each handles the nonlinearity in (14). In (15), the nonlinearity is handled by outer linearization, whereas (MISOCO 7) uses the constraints (47), (48), and:

$$\sum_{k=1}^{n} z_{jk} \leq \rho_j, \quad j = 1, \ldots, m.$$

3.4 Summary of MISOCO Formulations

We have shown how the service system design problem with congestion under the $M/M/1$ assumption can be cast as a MISOCO in seven different ways.

With the approaches presented in sections and, each SOCs are a function of t_j and s_j for a service center j, i.e., the stability constraint and the average delay in the queue respectively. In other words, in both approaches, the SOCs are ensuring a balance between the level of utilization of the service centers and the delay in the queue. That way, if the level of utilization grows, then t_j approaches zero and to ensure that the conic constraint satisfies the average delay s_j must increase, and the other way around. The difference between the two approaches is that in that balance is limiting total demand coming to the service center j, while it is limiting its service capacity. Notice that the first approach usually leads to SOCs with higher dimensions since there are usually more demand sources than service centers. For the two approaches, one can also find reformulations where the dimensions of the cones is fixed to three while increasing the number of variables in the model.

An alternative to using the stability constraint and the delay in the queue one can use the traffic intensity for each queue in the system. The traffic intensity is a ratio of the total demand coming to a service center to its service capacity. In this case, we obtain a set of SOCs that balance the delay in the queue and its traffic intensity. That balance is equivalent to the one described for the approaches in sections and specifically, if the traffic intensity approaches one, then the average delay in the queue increases, and when the traffic intensity approaches zero, then the average delay in the queue decreases. An interesting feature of using the traffic intensity is that the cones that model this balance involve only the variables s_j and ρ_j.

Table 1 Comparison of the MISOCO models

MISOCO model	1	2	3	4	5	6	7
# additional variables	$2m$	$(2+\ell)m$	$2m$	$2m$	$3m$	$2m$	$(2+\ell)m$
# SOCs	m	ℓm	m	mn	m	$2m$	m
Dimension of the SOCs	$2+\ell$	3	$2+n$	3	3	3 and $2+\ell$	3

We present a summary in Table 1 of how the dimensions of each of the seven models are affected by the choices made in each case. We compare their dimensions against the base model, (2)–(5).

The interest in this comparison comes from the possibility of using off-the-shelf solvers to solve these problems. For this reason, we performed some computational experiments to analyze the behavior of these different models and report the results in the next section.

4 Computational Testing

For our computational tests, we used the test sets from Holmberg et al. [19]. This set contains 71 different instances, with different numbers of demand and service centers. We modified these instances using the procedure described in [13]. Following that procedure, we increase the cost of a time unit in the system d each time by a factor of 10, starting with $d = 0.1$ until $d = 1000$, which resulted in a total of 355 instances. Also, we define the three different capacity levels as $\mu_{jk} = k\mu_j$, for $k = 1, 2, 3$ and $j = 1, \ldots, n$. The fixed cost is then defined as $p_{jk} = u_j^{\frac{2}{1+k}} \mu$, where $u_j = \frac{p_j}{\mu_j}$. Here, the values for p_j and μ_j are taken from the test sets in [19]. We use as a baseline our implementation of Elhedhli's method described in Sect. 2.2 for which we use a set of initial cuts obtained following the procedure in [12].

Because the formulations are not always ranked in the same order for different problems, we use a performance profile (PP) as proposed in [10]. The key for building a PP is $r_{p,s}$: the ratio between the time it took to solve problem p with formulation s and the best solution time to solve problem p over all formulations. In our PP, the vertical axis has the probability that $\log_2(r_{p,s}) \leq \tau$, where τ is the coordinate on the horizontal axis. For each level of τ, we consider a formulation successful if $\log_2(r_{p,s}) \leq \tau$. Hence, for each choice of τ, this gives the proportion of problems for which each formulation was successful. Our results are summarized in the PP in Fig. 2, which uses CPU time as the performance measure. We ran these experiments in a computer running linux centos with two Intel Xeon X5650 processors, each of which has 6 cores, and 36 GB or RAM. For these experiments, we set a time limit of 10,800 s and used CPLEX 12.6.3 to solve both the MISOCO problems and for our implementation of Elhedhli's method.

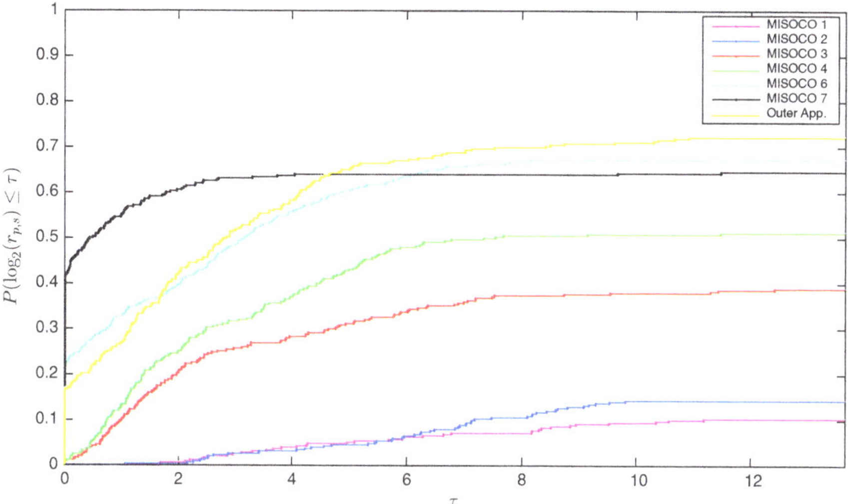

Fig. 2 Performance profile based in CPU time in seconds

Our results show that the probability of formulation MISOCO 7 being the fastest on a given problem is 0.41. The next best choice would be MISOCO 6 with a probability greater than 0.2 of being the fastest on a given problem, and third is Elhedhli's method with a probability of about 0.18 of being fastest on a given problem. For all the other formulations, the probability is below 0.05.

It turned out that the MILO and the MISOCO 5 reformulations were completely impractical because CPLEX was not successful in solving any of the instances with these models. In particular, MISOCO 5 is an interesting case because in terms of the dimensions shown in Table 1, its dimensions are comparable or even better than the other MISOCO formulations. However, we have observed that the solutions for the continuous relaxation of MISOCO 5 tend to have more fractional variables when compared with the other MISOCO formulations. This is in general a problem with a branch and bound since it results in bigger search trees. Our experiments point in the direction of the way the second and third constraints behave. Without providing any formal proof, our intuition is that squaring u_j allows to consider solutions with fractional y-s that are not feasible in the other formulations. Hence, for the sake of clarity, and due to the poor performance of MILO and MISOCO 5, we omitted them in the PP. A remark about the MILO approach presented in Sect. 2.1 is in place here. The MILO formulation always ended with an integrality gap greater than 5% within our time limit. We noted that the integer solution available at the moment we stopped CPLEX was in several cases very close to the optimal solution of the problem. This is an indicator that the linear relaxation of MILO reformulation is not very tight, which makes it difficult for the solver to improve the lower bound.

Notice that the choice between Elhedhli's method and MISOCO 6 becomes indifferent for $\tau \geq 2$, and also that these two options follow a similar trend. Recall that we are using the base 2 logarithm of the ratio $r_{p,s}$ to scale our performance profiles, hence this result translates to accepting being within a factor of 4 of the best solver to have Elhedhli's method match MISOCO 6 performance.

The gap between Elhedhli's method and MISOCO 7 is even more significant. Elhedhli's method catches up with the performance of MISOCO 7 only for $\tau > 4$. In other words, if the target is to be within a factor greater than 16 of the best approach, then any of MISOCO 7, MISOCO 6, or Elhedhli's method suffice. It is also clear from the PP that these three options dominate all the other formulations.

From our results, it is difficult to identify precisely the reason for the dominance of MISOCO 7. One thing that is observable in Table 1 is that MISOCO 7 is among those with the lowest number of SOCs, and all its cones have dimension 3. This is not the whole story though; for instance, MISOCO 6 has twice the number of cones of MISOCO 1, and the same number of cones with dimension $2 + \ell$, but still MISOCO 6 dominates MISOCO 1 (and in fact it dominates all formulations but MISOCO 7).

Finally, notice that the PP at the far right gives the results of the testing while ignoring the ratio comparison. We see that Elhedhli's method was able to solve more than 73% of the problems while MISOCO 6 and MISOCO 7 were able to solve 68% and 65% of the problems respectively. In summary, none of these three approaches was able to solve the full set of problems, and they showed a similar behavior in terms of scalability.

5 Conclusions

In this paper, we presented seven different MISOCO formulation for the service system design problem with congestion. Our computational results show that some of our conic formulations have a better performance when compared with existing exact approaches. This opens the possibility to use current off-the-shelf solvers like CPLEX, MOSEK, and GUROBI. Through the comparison of several conic formulations for the same problem, our experiments suggest that formulations with cones with small dimensions tend to dominate the formulations with cones of higher dimensions. This is observed when comparing MISOCO 1 with MISOCO 2, when comparing MISOCO 3 with MISOCO 4, and when comparing MISOCO 6 with MISOCO 7. Our results also show the importance of choosing the appropriate modeling technique. In particular, our experiments show that for obtaining reasonable solution times, the modeler should use MISOCO models 6 and 7.

An important feature of all these formulations is that they may be tackled with the recent disjunctive cuts approaches recently proposed in the literature. Given the encouraging results obtained with the MISOCO formulations presented in this work, our next step is to embed some of those cuts in a branch and cut algorithm to study its effect in the performance of the solvers.

Acknowledgements The first author acknowledges the support of the 2013 GERAD Postdoctoral Scholarship for the development of this research.

This work was supported by the Canada Research Chair in Discrete Nonlinear Optimization in Engineering.

References

1. Ahmadi-Javid, A., Hoseinpour, P.: Convexification of queueing formulas by mixed-integer second-order cone programming: An application to a discrete location problem with congestion (2017). arXiv:1710.05794
2. Amiri, A.: Solution procedures for the service system design problem. Comput. Oper. Res. **24**(1), 49–60 (1997)
3. Andersen, K., Jensen, A.: Intersection cuts for mixed integer conic quadratic sets. In: Goemans, M., Correa, J. (eds.) Integer Programming and Combinatorial Optimization. Lecture Notes in Computer Science, vol. 7801, pp. 37–48. Springer, Berlin, Heidelberg (2013)
4. Atamtürk, A., Narayanan, V.: Conic mixed-integer rounding cuts. Math. Program. **122**(1), 1–20 (2010)
5. Belotti, P., Góez, J., Pólik, I., Ralphs, T., Terlaky, T.: Disjunctive conic cuts for mixed integer second order cone optimization. Technical Report, G-2015-98, GERAD (Sept 2015)
6. Belotti, P., Góez, J.C., Pólik, I., Ralphs, T.K., Terlaky, T.: A conic representation of the convex hull of disjunctive sets and conic cuts for integer second order cone optimization. In: Al-Baali, M., Grandinetti, L., Purnama, A. (eds.) Numerical Analysis and Optimization: NAO-III, Muscat, Oman, January 2014, pp. 1–35. Springer International Publishing, Cham (2015)
7. Ben-Tal, A., Nemirovski, A.: Lectures on Modern Convex Optimization. Society for Industrial and Applied Mathematics (2001)
8. Berman, O., Krass, D.: Facility location problems with stochastic demands and congestion. In: Drezner, Z., Hamacher, H. (eds.) Facility Location: Applications and Theory, 1st edn., pp. 329–371. Springer-Verlag Berlin Heidelberg, New York, (2002)
9. Castillo, I., Ingolfsson, A., Sim, T.: Social optimal location of facilities with fixed servers, stochastic demand, and congestion. Prod. Oper. Manag. **18**(6), 721–736 (2009)
10. Dolan, D., Moré, J.: Benchmarking optimization software with performance profiles. Math. Program. **91**(2), 201–213 (2002)
11. Drewes, S.: Mixed integer second order cone programming. Ph.D. thesis, Technische Universität Darmstadt, Germany (2009)
12. Elhedhli, S.: Exact solution of a class of nonlinear knapsack problems. Oper. Res. Lett. **33**(6), 615–624 (2005)
13. Elhedhli, S.: Service system design with immobile servers, stochastic demand, and congestion. Manuf. Ser. Oper. Manag. **8**(1), 92–97 (2006)
14. Glover, F.: Improved linear integer programming formulations of nonlinear integer problems. Manag. Sci. **22**(4), 455–460 (1975)
15. Góez, J.: Mixed integer second order cone optimization disjunctive conic cuts: theory and experiments. Ph.D. Thesis, Lehigh University (2013)
16. Günlük, O., Linderoth, J.: Perspective reformulation and applications. In: Lee, J., Leyffer, S. (eds.) Mixed Integer Nonlinear Programming. The IMA Volumes in Mathematics and its Applications, vol. 154, pp. 61–89. Springer, New York (2012)
17. GUROBI: gurobi (2014). http://www.gurobi.com/resources/documentation
18. Hijazi, H., Bonami, P., Ouorou, A.: An outer-inner approximation for separable mixed-integer nonlinear programs. INFORMS J. Comput. **26**(1), 31–44 (2014)
19. Holmberg, K., Ronnqvist, M., Yuan, D.: An exact algorithm for the capacitated facility location problem with single sourcing. Eur. J. Oper. Res. **113**(3), 544–559 (1999). March

20. IBM: IBM ILOG CPLEX Optimization Studio V12.4 (2013). http://publib.boulder.ibm.com/infocenter/cosinfoc/v12r4/index.jsp
21. Keller, M., Karl, H.: Response time-optimized distributed cloud resource allocation. In: Proceedings of the 2014 ACM SIGCOMM Workshop on Distributed Cloud Computing, pp. 47–52. DCC '14. ACM, New York, NY, USA (2014)
22. Kılınç, M., Linderoth, J., Luedtke, J.: Effective separation of disjunctive cuts for convex mixed integer nonlinear programs. Optimization Online (2010)
23. Kılınç, M., Linderoth, J., Luedtke, J., Miller, A.: Strong-branching inequalities for convex mixed integer nonlinear programs. Comput. Optim. Appl. **59**(3), 639–665 (2014)
24. Kılınç-Karzan, F., Yıldız, S.: Two-term disjunctions on the second-order cone. In: Lee, J., Vygen, J. (eds.) Integer Programming and Combinatorial Optimization. Lecture Notes in Computer Science, vol. 8494, pp. 345–356. Springer International Publishing (2014)
25. Modaresi, S., Kılınç, M.R., Vielma, J.P.: Split cuts and extended formulations for mixed integer conic quadratic programming. Oper. Res. Lett. **43**(1), 10–15 (2015)
26. Modaresi, S., Kılınç, M.R., Vielma, J.P.: Intersection cuts for nonlinear integer programming: convexification techniques for structured sets. Math. Program. **155**(1), 575–611 (2016)
27. MOSEK: The MOSEK optimization tools manual, Version 7.0 (2013). http://mosek.com/resources/doc/
28. Torres, F.: Linearization of mixed-integer products. Math. Program. **49**(1–3), 427–428 (1990)
29. Vidyarthi, N., Jayaswal, S.: Efficient solution of a class of location-allocation problems with stochastic demand and congestion. Comput. Oper. Res. **48**, 20–30 (2014)
30. Wang, Q., Batta, R., Rump, C.M.: Algorithms for a facility location problem with stochastic customer demand and immobile servers. Ann. Oper. Res. **111**(1), 17–34 (2002)
31. Wang, Y.: Service system design with immobile servers, stochastic demand and economies of scale. Master's thesis, University of Waterloo (2015)
32. Zhang, Y., Berman, O., Verter, V.: Incorporating congestion in preventive healthcare facility network design. Eur. J. Oper. Res. **198**(3), 922–935 (2009)

The Iteration-Complexity Upper Bound for the Mizuno-Todd-Ye Predictor-Corrector Algorithm is Tight

Murat Mut and Tamás Terlaky

Abstract It is an open question whether there is an interior-point algorithm for linear optimization problems with a lower iteration-complexity than the classical bound $\mathcal{O}(\sqrt{n}\log(\frac{\mu_1}{\mu_0}))$. This paper provides a negative answer to that question for a variant of the Mizuno-Todd-Ye predictor-corrector algorithm. In fact, we prove that for any $\varepsilon > 0$, there is a redundant Klee-Minty cube for which the aforementioned algorithm requires $n^{(\frac{1}{2}-\varepsilon)}$ iterations to reduce the barrier parameter by at least a constant. This is provably the first case of an adaptive step interior-point algorithm where the classical iteration-complexity upper bound is shown to be tight.

Keywords Curvature · Central path · Polytopes · Complexity · Interior-point methods · Linear optimization

Mathematics Subject Classification (2000) 65K05 · 68Q25 · 90C05 · 90C51 90C60

1 Introduction

The paper of Karmarkar [5] in 1984 launched the field of interior-point methods (IPMs). Since then, IPMs have changed the landscape of optimization theory and been extended successfully for linear, nonlinear, and conic linear optimization [12].

For linear optimization problems (LO), to reduce the barrier parameter from μ_1 to μ_0, the best known iteration-complexity upper bound is $\mathcal{O}(\sqrt{n}\log(\frac{\mu_1}{\mu_0}))$. In practice however, IPMs require much less iterations than predicted by the theory. It has been conjectured that the required number of iterations grows logarithmically in the

M. Mut (✉) · T. Terlaky
Department of Industrial and Systems Engineering, Lehigh University, Bethlehem, PA, USA
e-mail: mutmurat2@gmail.com

T. Terlaky
e-mail: terlaky@lehigh.edu

J. D. Pintér and T. Terlaky (eds.), *Modeling and Optimization: Theory and Applications*, Springer Proceedings in Mathematics & Statistics 279,
https://doi.org/10.1007/978-3-030-12119-8_6

number of variables [4]. Sonnevend et al. [15] showed that for two distinct special classes of LO problems, we have the complexity upper bounds $\mathcal{O}(n^{\frac{1}{4}}\log(\frac{\mu_1}{\mu_0}))$ and $\mathcal{O}(n^{\frac{3}{8}}\log(\frac{\mu_1}{\mu_0}))$. Using an "anticipated" iteration-complexity analysis, [7] gives an $\mathcal{O}(n^{\frac{1}{4}}\log(\frac{\mu_1}{\mu_0}))$ iteration-complexity bound for certain path-following IPM algorithms. Huhn and Borgwardt [3] present a thorough probabilistic analysis of the iteration-complexity of IPMs and establish that under the rotation-symmetry model, average iteration-complexity is strongly polynomial.

Another direction of research regarding the iteration-complexity of IPMs is to construct worst-case examples. Sonnevend et al. [15] established a lower bound for a predictor-corrector type algorithm for LO problems. This algorithm is very similar to the predictor-corrector algorithm of [8], which has later become known as the Mizuno-Todd-Ye predictor-corrector algorithm (MTY P-C). The variant by Sonnevend et al. [15], which is also the main focus of our paper originated in [15, 16]. The algorithm in [15] requires $\Omega(n^{\frac{1}{3}})$ iterations to reduce the duality gap by $\log n$ for certain LO problems. In this paper, due to their similarity, we will refer to both the Mizuno-Todd-Ye predictor-corrector algorithm and its earlier variant by Sonnevend et al. [15] as MTY P-C algorithm. In Sect. 2, we summarize and highlight their difference. A similar worst-case lower bound of $\Omega(n^{\frac{1}{3}})$ for another LO problem construction was established by Todd et al. [17]. They proved that the primal-dual affine scaling algorithm takes $\Omega(n^{\frac{1}{3}})$ iterations to reduce the duality gap by a constant. This result has later been extended by Todd and Ye [18] for long step primal-dual IPMs.

In a series of papers [1, 2, 10, 11], LO problems have been constructed with central paths making a large number of sharp turns with the intuitive idea that for a path-following algorithm each turn should lead to an extra Newton step. These constructions share the common feature; that is, the (dual) feasible set is a perturbed Klee-Minty (KM) cube and the central path visits all the vertices of the KM cube. In [11], for instance, the authors show that the central path makes $\Omega(\frac{\sqrt{n}}{\sqrt{\log n}})$ sharp turns.

A curvature integral developed by [15, 16] accurately estimates the number of iterations of a variant of MTY P-C algorithm, see Sect. 2. This curvature integral is one of the main tools in our paper and we will refer to this curvature as *Sonnevend's curvature*.

In this paper, we build our work upon the KM construction in [11]. The main argument of the paper can be summarized as follows: We first prove that a KM construction [11] with a carefully chosen neighborhood of the central path, which depends on the dimension of the cube visits every vertices of the cube in such a way that following the central path within that neighborhood requires an exponential number of steps. From Theorem 2.1, this yields a large lower bound for the Sonnevend's curvature. Then by using a modified hybrid version of that construction as well as Theorem 2.1 once again, we are able to conclude that for any $\varepsilon > 0$, there is a redundant hybrid version of the KM cube for which the MTY P-C algorithm requires $\Omega\left(n^{(\frac{1}{2}-\varepsilon)}\log\left(\frac{\mu_1}{\mu_0}\right)\right)$ where $\log\frac{\mu_1}{\mu_0} = \mathcal{O}(\log n)$. Hence by a rigorous analysis, our modified KM construction provides the first case of an IPM, the MTY P-C algorithm, for which the classical iteration-complexity upper bound is essentially tight.

In the rest of this section, the basic terminology used in this paper is presented.

Let A be an $m \times n$ matrix of full rank. For $c \in \mathbb{R}^n$ and $b \in \mathbb{R}^m$, we consider the standard form primal and dual linear optimization problems,

$$\begin{array}{llll} \min c^T x & & \max & b^T y \\ \text{s.t.} & Ax = b & \text{s.t.} & A^T y + s = c \\ & x \geq 0, & & s \geq 0, \end{array} \tag{1}$$

where $x, s \in \mathbb{R}^n$, $y \in \mathbb{R}^m$ are vectors of variables. Denote the sets of primal and dual feasible solutions by $\mathcal{P} = \{x \in \mathbb{R}^n : Ax = b,\ x \geq 0\}$ and $\mathcal{D} = \{(y, s) \in \mathbb{R}^m \times \mathbb{R}^n : A^T y + s = c,\ s \geq 0\}$; the sets of strictly feasible primal and dual solutions by $\mathcal{P}^+$ and $\mathcal{D}^+$, respectively. Without loss of generality, see e.g., [14], we may assume that $\mathcal{P}^+ \neq \emptyset$ and $\mathcal{D}^+ \neq \emptyset$. For a parameter $\mu > 0$ and a vector $w > 0$, the w-weighted path equations are given by

$$\begin{aligned} Ax = b,&\quad x \geq 0 \\ A^T y + s = c,&\quad s \geq 0 \\ xs = \mu w,& \end{aligned} \tag{2}$$

where uv denotes $[u_1 v_1, \ldots, u_n v_n]^T$ for $u, v \in \mathbb{R}^n$. For $w = e$, with e being the all-one vector, Eq. (2) gives the central path equations.

2 IPMs and Sonnevend's Curvature of the Central Path

First, we briefly review the relevant algorithms to this paper.

Roughly speaking, path-following IPMs differ by the way the barrier parameter $\mu^+ := (1 - \theta)\mu$ is chosen and for what values of μ, the Newton steps are calculated. While for short-step IPMs, we have $\theta = \Omega(\frac{1}{\sqrt{n}})$, predictor-corrector type algorithms allow a larger θ, hence a larger reduction in μ. Given $\mu > 0$, and $\beta > 0$, we define the β-neighborhood of the point on the central path corresponding to μ as

$$\mathcal{N}(\beta, \mu) := \{(x, s) \in \mathcal{P}^+ \times \mathcal{D}^+ : \left\| \frac{xs}{\mu} - e \right\| \leq \beta\}. \tag{3}$$

The β-neighborhood of the central path is defined as $\mathcal{N}(\beta) := \bigcup_{\mu > 0} \mathcal{N}(\beta, \mu)$.

Both the algorithm of [16] and the MTY P-C algorithm use two nested neighborhoods $\mathcal{N}(\beta_0)$ and $\mathcal{N}(\beta_1)$ for $0 < \beta_0 < \beta_1 < 1$. The MTY P-C algorithm alternates between two search directions: The predictor search direction is used within the smaller neighborhood $\mathcal{N}(\beta_0)$ and it aims to reduce μ to zero. Let (x, s) be the current iterate, $(\Delta x, \Delta s)$ the predictor search direction and $(x^+, s^+) :=$

$(x + \theta \Delta x, s + \theta \Delta s)$. The MTY P-C algorithm and the algorithm in [16] differ in the way the value of θ is determined. In the MTY P-C algorithm, θ is determined as being the largest step for which (x^+, s^+) stays within the larger neighborhood $\mathcal{N}(\beta_1)$. In the algorithm of [16], the value of θ is determined as the largest number for which $\left\| \frac{x^+ s^+}{\mu^+} - \xi \right\| \leq \beta_1$, where $\xi = \frac{xs}{\mu}$. Then a pure centering step is taken which will take the iterate back to the smaller neighborhood $\mathcal{N}(\beta_0)$ in such a way that the normalized duality gap $\mu = \frac{x^T s}{n}$ does not change. Both algorithms can take long steps, in fact, it is known that [13, 16] as $k \to \infty$, $\theta_k \to 1$, where θ_k is the step length of the predictor direction at iteration k.

Sonnevend's curvature, introduced in [15], is closely related to the iteration-complexity of a variant of the MTY P-C algorithm. Let $\kappa(\mu) = \|\mu \dot{x} \dot{s}\|^{1/2}$. Stoer et al. [16] proved that their predictor-corrector algorithm has a complexity bound, which can be expressed in terms of $\kappa(\mu)$.

Theorem 2.1 ([16]) *Let the nested neighborhood parameters β_0, β_1 of the MTY P-C algorithm satisfy $\beta_0 + \beta_1 \leq \frac{1}{2}$. Let N be the number of iterations of the MTY P-C algorithm to reduce the barrier parameter from μ_1 to μ_0. Suppose $\kappa(\mu) \geq \nu$ for some constant $\nu > 0$ on $\mu \in [\mu_0, \mu_1]$. Then for some "universal" constants C_1 and C_2 that depend only on the neighborhood of the central path, we have*

$$C_3 \int_{\mu_0}^{\mu_1} \frac{\kappa(\mu)}{\mu} d\mu - 1 \leq N \leq C_1 \int_{\mu_0}^{\mu_1} \frac{\kappa(\mu)}{\mu} d\mu + C_2 \log \left(\frac{\mu_1}{\mu_0} \right) + 2. \tag{4}$$

Constant C_3 depends on ν as well as the neighborhood of the central path.

The following proposition states the basic properties of Sonnevend's curvature.

Proposition 2.1 ([15]) *The following holds.*

1. *We have* $\kappa(\mu) = \left\| \frac{\mu \dot{s}(\mu)}{s(\mu)} - \left(\frac{\mu \dot{s}(\mu)}{s(\mu)} \right)^2 \right\|^{\frac{1}{2}}$.
2. *We have* $\left\| \frac{\mu \dot{s}(\mu)}{s(\mu)} \right\| \leq \sqrt{n}$ *and* $\kappa(\mu) \leq \sqrt{n}$ *implying that*

$$\int_{\mu_0}^{\mu_1} \frac{\kappa(\mu)}{\mu} d\mu = \mathcal{O} \left(\sqrt{n} \log \left(\frac{\mu_1}{\mu_0} \right) \right).$$

3 KM Cube Construction

First, we recall the KM construction in [11] and review its fundamental properties. First, consider the following squashed unit cube $[0, 1]^m$ with a factor of $\rho < \frac{1}{2}$.

$$\begin{array}{lrl} \max & -y_m & \\ \text{s.t.} & 0 \leq y_1 \leq 1 & \\ & \rho y_{k-1} \leq y_k \leq 1 - \rho y_{k-1} & \text{for } k = 2, \ldots, m. \end{array} \tag{5}$$

The problem (5) has $2m$ constraints, and m variables. Certain variants of the simplex method take $2^m - 1$ to solve this problem, e.g., see [6] Sect. 3. The simplex path for these variants starts from $(y_1, \ldots, y_m) = (0, \ldots, 0, 1)$, and it visits all the vertices ordered by the decreasing value of the last coordinate y_m until reaching the optimal point, which is the origin.

Next, we consider redundant constraints added to the formulation (5) as follows. Note that we formulated (6) as the dual problem in (1) and $b^T = [0, \ldots, 0, -1]$ and c has coordinates from the set $\{0, 1, d_1, \ldots, d_m\}$.

$$\begin{array}{lrl} \max & -y_m & \\ \text{s.t.} & 0 \leq y_1 \leq 1 & \\ & \rho y_{k-1} \leq y_k \leq 1 - \rho y_{k-1} & \text{for } k = 2, \ldots, m. \\ & 0 \leq d_1 + y_1 & \text{repeated } h_1 \text{ times} \\ & 0 \leq d_2 + y_2 & \text{repeated } h_2 \text{ times} \\ & \cdots & \\ & 0 \leq d_m + y_m & \text{repeated } h_m \text{ times.} \end{array} \tag{6}$$

As in [11], we fix

$$\rho(m) := \frac{m}{2(m+1)} \text{ and } d := \left(\frac{1}{\sqrt{\rho(m)^{m-1}}}, \frac{1}{\sqrt{\rho(m)^{m-2}}}, \ldots, \frac{1}{\sqrt{\rho(m)}}, 0 \right). \tag{7}$$

Throughout the paper, let n be the number of inequalities in (6) with $n = h_1 + \cdots + h_m + 2m$. Note that the feasible regions of (5) and (6) are the same. We denote the m-dimensional KM cube by $\mathcal{KM}(m, \rho(m))$ for those regions. See Fig. 1 for $\mathcal{KM}(m, \rho(m))$ with $m = 2$.

In order to define a simplex path, we follow the encoding in [6] of the vertices of $\mathcal{KM}(m, \rho(m))$ with (0, 1) vectors. For (6), let the slack variables $\bar{s}_k = 1 - \rho(m) y_{k-1} - y_k$ and $s_k = y_k - \rho(m) y_{k-1}$ for $k = 2, \ldots, m$ with the convention $\bar{s}_1 = 1 - y_1$ and $s_1 = y_1$. There is a one-to-one correspondence between the vertices of $\mathcal{KM}(m, \rho(m))$ with the m-tuples $v^i \in \{0, 1\}^m$, $i = 1, \ldots, 2^m$ as follows. Each vertex of $\mathcal{KM}(m, \rho(m))$ is determined by whether exactly one of $s_i = 0$ or $\bar{s}_i = 0$ for each $i = 1, \ldots, m$ in (6). If $s_i = 0$, the ith coordinate of the corresponding m-tuple in $\{0, 1\}^m$ is 0; if $\bar{s}_i = 1$, it is 1. For our purpose, we describe the relevant terms of $\mathcal{KM}(m, \rho(m))$ inductively as follows:

First, we describe an ordering of the set of the vertices $\mathcal{V}(m)$ of $\mathcal{KM}(m, \rho(m))$, [6]. Note that $\mathcal{V}(m)$ is an encoding of the vertices of $\mathcal{KM}(m, \rho(m))$, they are not the actual vertex points in $\mathbb{R}^m$. For $m = 2$, let

$$\mathcal{V}(2) = \{v^1, v^2, v^3, v^4\} = \{(0, 1), (1, 1), (1, 0), (0, 0)\}. \tag{8}$$

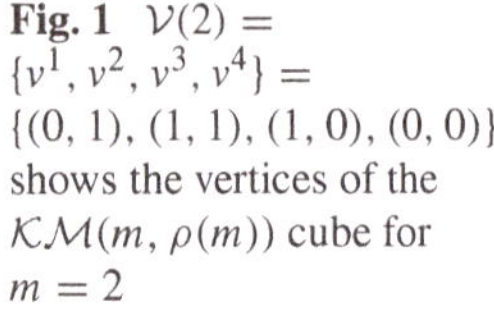

Fig. 1 $\mathcal{V}(2) = \{v^1, v^2, v^3, v^4\} = \{(0, 1), (1, 1), (1, 0), (0, 0)\}$ shows the vertices of the $\mathcal{KM}(m, \rho(m))$ cube for $m = 2$

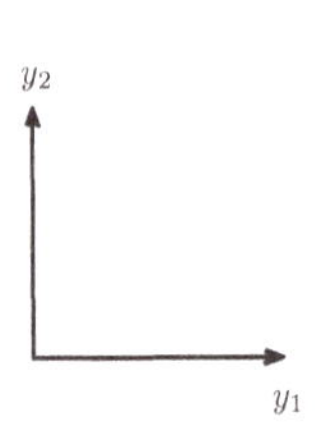

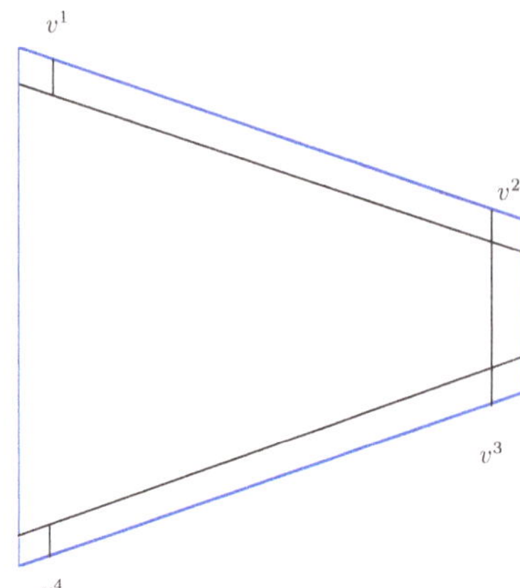

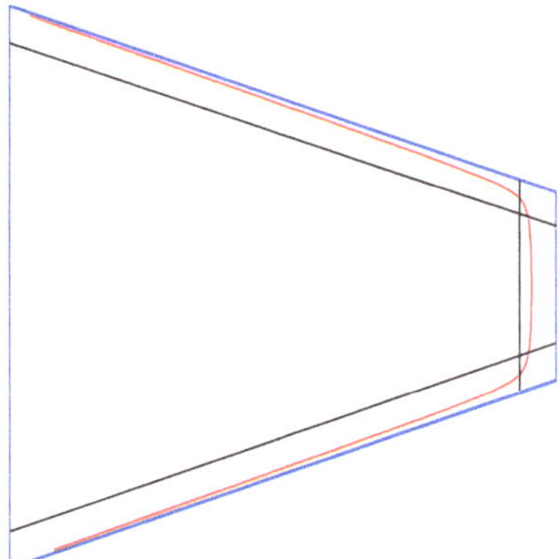

Fig. 2 The central path visits the vertices $\mathcal{V}(2) = \{v^1, v^2, v^3, v^4\}$ of the $\mathcal{KM}(m, \rho(m))$ cube for $m = 2$ in the given order as μ decreases

Figure 1 shows the vertices of the $\mathcal{KM}(m, \rho(m))$. Then let

$$\mathcal{V}(m+1) = \{(v^{2^m}, 1), (v^{2^m-1}, 1), \dots, (v^1, 1), (v^1, 0), (v^2, 0), \dots, (v^{2^m}, 0)\}. \quad (9)$$

Then the simplex path visits 2^m vertices $\mathcal{V}(m)$ of $\mathcal{KM}(m, \rho(m))$ in the ordering in $\mathcal{V}(m)$, [6].

Next we define the central path $\mathcal{CP}(m)$ for the problem (6). In [11], it is shown that the central path $\mathcal{CP}(m)$ of (6) with a certain number of redundant constraints $(h_1, \dots, h_m)$ closely traces the simplex path in the order given in the set $\mathcal{V}(m)$. Figures 2 and 3 show the central path for $m = 2$ and $m = 3$.

Next we define inductively a tube along the edges of the simplex path in $\mathcal{KM}(m, \rho(m))$ as follows. Let $\delta \leq \frac{1}{4(m+1)}$. Let $\mathcal{T}_\delta^U(2) = \{y : \mathbb{R}^2 : \bar{s}_2 \leq \delta\}$, $\mathcal{T}_\delta^L(2) = \{y : \mathbb{R}^2 : s_2 \leq \delta\}$. Note that $\mathcal{T}_\delta^U(2)$ and $\mathcal{T}_\delta^L(2)$ corresponds to a tube for the upper and lower facets of $\mathcal{KM}(2, \rho(2))$ respectively. By $\mathcal{T}_\delta(2)$, denote the union $\mathcal{T}_\delta^L(2) \cup \mathcal{T}_\delta^U(2)$. Then for $m \geq 2$, define $\mathcal{T}_\delta^U(m+1) = \{y : \mathbb{R}^{m+1} : \bar{s}_{m+1} \leq \delta, (y_1, \dots, y_m) \in \mathcal{T}_\delta(m)\}$ and $\mathcal{T}_\delta^L(m+1) = \{y : \mathbb{R}^{m+1} : s_{m+1} \leq \delta, (y_1, \dots, y_m) \in \mathcal{T}_\delta(m)\}$. Notice that

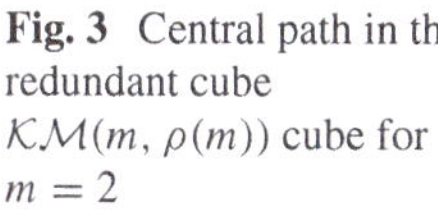

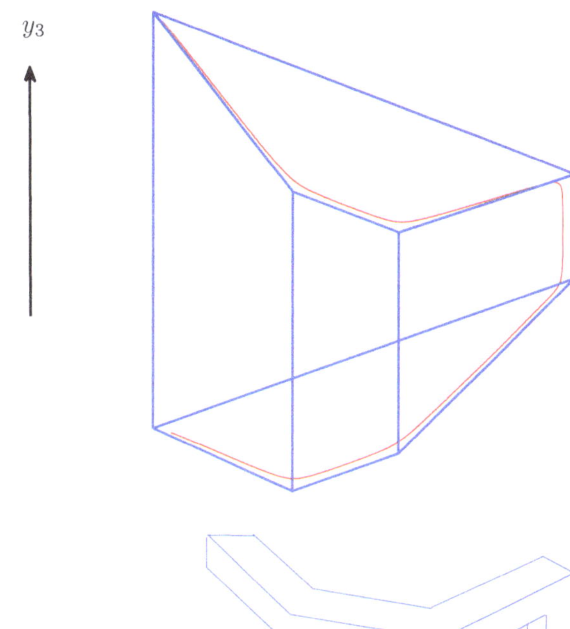

Fig. 3 Central path in the redundant cube $\mathcal{KM}(m, \rho(m))$ cube for $m = 2$

Fig. 4 Illustration of the tube $\mathcal{T}_\delta(m)$ for $m = 3$

$\mathcal{T}_\delta^U(3)$ is a tube that corresponds to the upper facet of $\mathcal{KM}(3, \rho(3))$ where $y_3 = 1 - \rho(3)y_2$. Similarly $\mathcal{T}_\delta^L(3)$ is a tube that corresponds to the lower facet of $\mathcal{KM}(3, \rho(3))$ where $y_3 = \rho(3)y_2$. Also these upper and lower facets are $\mathcal{KM}(2, \rho(3))$ cubes themselves, see Fig. 3. Finally for a general $m \geq 2$, define $\mathcal{T}_\delta(m) = \mathcal{T}_\delta^L(m) \cup \mathcal{T}_\delta^U(m)$. Hence inside inside $\mathcal{KM}((m+1), \rho(m+1))$, by identifying the projection of $(y_1, \ldots, y_m, y_{m+1})$ onto $(y_1, \ldots, y_m) \in \mathcal{KM}(m, \rho(m+1))$, and considering the assumption that δ is decreasing in m, we can write $\mathcal{T}_\delta^U(m+1) \subset \mathcal{T}_\delta(m)$ and $\mathcal{T}_\delta^L(m+1) \subset \mathcal{T}_\delta(m)$, see Fig. 4.

Recall that each vertex of $\mathcal{KM}(m, \rho(m))$ is determined by whether exactly one of $s_i = 0$ or $\bar{s}_i = 0$ for each $i = 1, \ldots, m$. In a similar fashion, we now define a δ-neighborhood of a vertex of $\mathcal{KM}(m, \rho(m))$ by whether exactly one of $s_i \leq \delta$ or $\bar{s}_i \leq \delta$ for each $i = 1, \ldots, m$ in (6). Figure 1 displays the δ-neighborhoods of the vertices $\mathcal{V}(2) = \{v^1, v^2, v^3, v^4\}$ of the $\mathcal{KM}(m, \rho(m))$ cube for $m = 2$.

The following proposition is essentially Proposition 2.2 in [11].

Proposition 3.1 *In (6), one can choose the parameters in such a way that the central path $\mathcal{CP}(m)$ in $\mathcal{KM}(m, \rho(m))$ stay inside the tube $\mathcal{T}_\delta(m)$. In particular, one can choose $\rho(m) = \frac{m}{2(m+1)}$, $\delta \leq \frac{1}{4(m+1)}$ so that $n = \mathcal{O}(m2^{2m})$. As μ decreases, the central path visits the δ-neighborhoods of the vertices given in the order by (9). Moreover, the number of inequalities n is linear in $\frac{1}{\delta}$.*

Proof See Proposition 2.2 in [11]. □

Now for $\mathcal{KM}(m, \rho(m))$, we identify two regions R_δ^U and R_δ^L within tube $\mathcal{T}_\delta(m)$ in such a way that going from R_δ^U to R_δ^L (an vice versa) with line segments staying inside tube $\mathcal{T}_\delta(m)$ requires $\Omega(2^{m-1})$ number of iterations. Let

$$R_\delta^U := \{y \in \mathcal{KM}(m, \rho(m)) : s_1 \leq \delta,\ s_2 \leq \delta, \ldots, s_{m-1} \leq \delta,\ \bar{s}_m \leq \delta\} \tag{10}$$

and

$$R_\delta^L := \{y \in \mathcal{KM}(m, \rho(m)) : s_1 \leq \delta,\ s_2 \leq \delta, \ldots, s_{m-1} \leq \delta,\ s_m \leq \delta\}. \tag{11}$$

We have the following.

Proposition 3.2 *For $\mathcal{KM}(m, \rho(m))$, let $y^U \in R_\delta^U$ and $y^L \in R_\delta^L$. Then staying inside the tube $\mathcal{T}_\delta(m)$, one requires at least 2^{m-1} line segments to reach y^U from y^L and vice versa.*

Proof With the parameters chosen as in Proposition 3.1, we first show $\mathcal{T}_\delta^U(m)$ and $\mathcal{T}_\delta^L(m)$ do not intersect for any m. Suppose by contradiction that there is a $y \in \mathcal{T}_\delta^U(m) \cap \mathcal{T}_\delta^L(m)$. From the definition of $\mathcal{T}_\delta^U(m)$ and $\mathcal{T}_\delta^L(m)$, we have $\bar{s}_m = 1 - \rho(m)y_{m-1} - y_m \leq \delta$ and $s_m = y_m - \rho(m)y_{m-1} \leq \delta$. Adding these two inequalities, we get $1 - 2\rho(m)y_{m-1} \leq 2\delta$. By the choice of $\rho(m)$ and δ, it is easy to see that this will lead to the contradiction $y_{m-1} > 1$. Hence $\mathcal{T}_\delta^U(m) \cap \mathcal{T}_\delta^L(m) = \emptyset$.

The rest of the proof is by induction on m. For $m = 2$, let $y^U \in R_\delta^U$ and $y^L \in R_\delta^L$ with $\delta \leq \frac{1}{4(m+1)}$. Then, for y^U we have $s_1 = y_1 \leq \delta$ and $\bar{s}_2 \leq \delta$ which implies that $y_2 \geq 1 - \delta - \rho(2)\delta \geq 1 - 2\delta = \frac{5}{6}$. Analogously, for y^L, we have $s_1 = y_1 \leq \delta$ and $s_2 \leq \delta$ which implies $y_2 \leq \delta + \rho(2)y_1 \leq 2\delta = \frac{1}{6}$. Clearly, staying inside the tube $\mathcal{T}_\delta(2)$, it takes at least 2 iterations to reach a point with $y_2 \leq \frac{1}{6}$ from a point with $y_2 \geq \frac{5}{6}$, see Fig. 1.

As inductive step, suppose that to reach any point in R_δ^L from a point in R_δ^U with $R_\delta^L \subset \mathcal{KM}(m-1, \rho(m-1))$ and $R_\delta^U \subset \mathcal{KM}(m-1, \rho(m-1))$, one requires at least 2^{m-2} steps with line segments staying inside $\mathcal{T}_\delta(m-1)$. Let $y^U \in R_\delta^U$ and $y^L \in R_\delta^L$ inside $\mathcal{T}_\delta(m) \subset \mathcal{KM}(m, \rho(m))$. We distinguish two points p^1 and p^2 such that

$$p^1 \in \{y \in \mathcal{KM}(m, \rho(m)) : s_1 \leq \delta,\ s_2 \leq \delta, \ldots, \bar{s}_{m-1} \leq \delta,\ \bar{s}_m \leq \delta\}$$

and

$$p^2 \in \{y \in \mathcal{KM}(m, \rho(m)) : s_1 \leq \delta,\ s_2 \leq \delta, \ldots, \bar{s}_{m-1} \leq \delta,\ s_m \leq \delta\}.$$

Note that the point p^1 belongs to the δ-neighborhood of the vertex $v^{2^{m-1}} = (0, 0, \ldots, 0, 1, 1)$ and the point p^2 belongs to the δ-neighborhood of the vertex point $v^{2^{m-1}+1} = (0, 0, \ldots, 0, 1, 0)$. Then, using the inductive definition of $\mathcal{T}_\delta^U(m)$ and $\mathcal{T}_\delta^L(m)$, it is easy to see that $y^U, p^1 \in \mathcal{T}_\delta^U(m)$ and $p^2, y^L \in \mathcal{T}_\delta^L(m)$. By inductive hypothesis, one needs at least 2^{m-2} line segments to reach p^1 from y^U staying inside the tube $\mathcal{T}_\delta^U(m) \subset \mathcal{T}_\delta(m-1)$. Similarly one needs at least 2^{m-2} line segments to reach y^L from p^2 staying inside the tube $\mathcal{T}_\delta^L(m) \subset \mathcal{T}_\delta(m-1)$. Moreover since by the first part of the proof, we have $\mathcal{T}_\delta^U(m) \cap \mathcal{T}_\delta^L(m) = \emptyset$, it follows that to reach y^L from y^U, one needs to traverse within $\mathcal{T}_\delta(m-1)$ twice, each time requiring at least 2^{m-2} steps. This proves that one requires at least 2^{m-1} line segments to reach y^U from y^L, hence the proof is complete. □

4 Neighborhood of the KM Cube Central Path

In Sect. 3, we showed that with $n = \mathcal{O}(m2^{2m})$ redundant constraints, the central path $\mathcal{CP}(m)$ stays inside a tube $\mathcal{T}_\delta(m)$. Moreover, we proved that it will take at least 2^{m-1} line segments to reach a point in R_δ^L close to the optimal solution of (6) from a point in R_δ^U close to the analytic center of $\mathcal{KM}(m, \rho(m))$. However, path-following IPMs algorithms including the MTY P-C algorithm, use the neighborhood $\mathcal{N}(\beta)$ as opposed to the tube neighborhood $\mathcal{T}_\delta(m)$ we used in Sect. 3. In this section, we analyze the $\mathcal{N}(\beta)$ neighborhood for the cube $\mathcal{KM}(m, \rho(m))$ and prove that for $\beta = \Omega(\frac{1}{m+1})$, we have $\mathcal{N}(\beta) \subset \mathcal{T}_\delta(m)$. In other words, with appropriately chosen neighborhood parameters of $\mathcal{KM}(m, \rho(m))$, all the iterates of the MTY P-C algorithm stay inside the tube $\mathcal{T}_\delta(m)$. Hence, we can draw the conclusion that for $\mathcal{KM}(m, \rho(m))$, the MTY P-C algorithm will require $\Omega(2^{m-1})$ iterations with the neighborhood $\mathcal{N}(\beta)$, where $\beta = \Omega(\frac{1}{m+1})$.

In order to find the largest β for which $\mathcal{N}(\beta) \subset \mathcal{T}_\delta(m)$, we will use weighted paths. The following lemma is essentially Lemma 4.1 in [16].

Lemma 4.1 *Fix μ and let $w > 0$ such that $\|w - e\| \leq \varepsilon$. Let $(x(w), y(w), s(w))$ denote the point on the w-weighted path which is the solution set of (2). Let $\Delta s_i = s_i(w) - s_i$, where the s_i values are the coordinates of the central path point for the fixed μ for $i = 1, \ldots, n$. Then we have $\left|\frac{\Delta s_i}{s_i}\right| \leq 2\varepsilon$ for $i = 1, \ldots, n$.*

When we apply the information in Lemma 4.1 to $\mathcal{KM}(m, \rho(m))$, we obtain the following result.

Lemma 4.2 *There exists a $\mathcal{KM}(m, \rho(m))$ with $n = \mathcal{O}(m2^{2m})$ such that all the w-weighted paths with $\|w - e\| \leq \beta := \frac{\delta}{4}$ stay inside the tube $\mathcal{T}_\delta(m)$ with $\delta \leq \frac{1}{4(m+1)}$.*

Proof Let $\delta \leq \frac{1}{4(m+1)}$. Then, from Proposition 3.1, we know that there exists $\mathcal{KM}(m, \rho(m))$ with $n = \mathcal{O}(m2^{2m})$ so that the central path stays inside the tube $\mathcal{T}_{\frac{\delta}{2}}(m)$. Choose $\beta = \frac{\delta}{4}$ for $\mathcal{KM}(m, \rho(m))$ so that $\|w - e\| \leq \beta$. Since for all the slacks, we have $s_i \leq 1$ or $\bar{s}_i \leq 1$, Lemma 4.1 implies that $s_i(w) \leq s_i + \frac{\delta}{2}$ and $\bar{s}_i(w) \leq \bar{s}_i + \frac{\delta}{2}$. Then whenever $s_i \leq \frac{\delta}{2}$ or $\bar{s}_i \leq \frac{\delta}{2}$, we have $\bar{s}_i(w) \leq \delta$ and $s_i(w) \leq \delta$. Since a tube $\mathcal{T}_\delta(m)$ with a general δ inside $\mathcal{KM}(m, \rho(m))$ is determined by these slacks, it follows that all w-weighted paths stay inside the tube $\mathcal{T}_\delta(m)$ with $\delta \leq \frac{1}{4(m+1)}$. This concludes the proof. □

Next lemma proves a result analogous to Lemma 4.2 tailored for R^U_δ and R^L_δ.

Lemma 4.3 *Let $\delta \leq \frac{1}{4(m+1)}$ and fix $\beta := \frac{\delta}{4}$. Suppose that $y(\mu_1) \in R^U_{\delta/2}$ for some μ_1. Then $\mathcal{N}(\beta, \mu_1) \subset R^U_\delta$. Similarly if for some μ_0, $y(\mu_0) \in R^L_{\delta/2}$, then $\mathcal{N}(\beta, \mu_0) \subset R^L_\delta$.*

Proof Suppose that for some μ_1, $y(\mu_1) \in R^U_{\delta/2}$, i.e., $s_1 \leq \frac{\delta}{2}$, $s_2 \leq \frac{\delta}{2}, \ldots, s_{m-1} \leq \frac{\delta}{2}, \bar{s}_m \leq \frac{\delta}{2}$. Let $y \in \mathcal{N}(\beta, \mu_1)$. Then, for $w := \frac{xs}{\mu_1}$, we have $\|w - e\| \leq \beta$. Since for all the slacks in $\mathcal{KM}(m, \rho(m))$, we have $s_i \leq 1$ or $\bar{s}_i \leq 1$, Lemma 4.1 implies that $s_i(w) \leq s_i + \frac{\delta}{2}$ and $\bar{s}_i(w) \leq \bar{s}_i + \frac{\delta}{2}$. Then whenever $s_i \leq \frac{\delta}{2}$ or $\bar{s}_i \leq \frac{\delta}{2}$, we have $\bar{s}_i(w) \leq \delta$ and $s_i(w) \leq \delta$. This proves $y \in R^U_\delta$, which implies $\mathcal{N}(\beta, \mu_1) \subset R^U_\delta$. The proof of the rest of the claim is similar. □

In the rest of this section, we aim to find an interval $[\mu_0, \mu_1]$ and an upper bound for $\log(\frac{\mu_1}{\mu_0})$ such that the neighborhoods $\mathcal{N}(\beta, \mu_1) \subset R^U_\delta$ and $\mathcal{N}(\beta, \mu_0) \subset R^L_\delta$ for some δ and β.

Let $\delta \leq \frac{1}{4(m+1)}$ and $(y_1(\mu_1), \ldots, y_m(\mu_1))$ be a central path $\mathcal{CP}(m)$ point such that $s_1 = \frac{\delta}{2}$, $s_2 \leq \frac{\delta}{2}, \ldots, \bar{s}_m \leq \frac{\delta}{2}$. Note that any point satisfying $s_1 = \frac{\delta}{2}$, $s_2 \leq \frac{\delta}{2}, \ldots, \bar{s}_m \leq \frac{\delta}{2}$ is inside the $\frac{\delta}{2}$-neighborhood of the vertex point $(0, 0, \ldots, 0, 1)$, hence Proposition 3.1 guarantees the existence of a central path point $(y_1(\mu_1), \ldots, y_m(\mu_1))$. Since for $((y_1(\mu_1), \ldots, y_m(\mu_1)))$, we have $s_1 = \frac{\delta}{2}$, $s_2 \leq \frac{\delta}{2}, \ldots, \bar{s}_m \leq \frac{\delta}{2}$ by assumption, we conclude $y(\mu_1) \in R^U_{\delta/2}$ from the definition of R^U_δ in (10). Then, by using Theorem 3.7 in [11], one can show that $\mu_1 \leq \frac{\rho(m)^{m-1}\delta}{2}$. Let us fix $\mu_1 = \frac{\rho(m)^{m-1}\delta}{2}$ and let $\beta := \frac{\delta}{4}$. Then Lemma 4.3 applied to $y(\mu_1)$ with the choice of μ_1 and β implies that the neighborhood $\mathcal{N}(\beta, \mu_1)$ stays inside the region R^U_δ. Hence, any point inside the neighborhood $\mathcal{N}(\beta, \mu_1)$ also stays inside the region R^U_δ.

Next, we will find a μ_0 such that the neighborhood $\mathcal{N}(\beta, \mu_0)$ is within the region R^L_δ. Let $(y_1(\mu_0), \ldots, y_m(\mu_0))$ be the central path point such that $y_m(\mu_0) = \frac{\rho(m)^{m-1}\delta}{2}$. Note that since the objective function in (6) is $-y_m$, a central point satisfying $y_m(\mu) = \frac{\rho(m)^{m-1}\delta}{2}$ exists and is unique. Since from (6), we have $\rho(m)y_i \leq y_{i+1}$ for $i = 1, \ldots, (m-1)$, we obtain $y_1(\mu) \leq \frac{\delta}{2}$, $y_2(\mu) \leq \frac{\delta}{2}, \ldots, y_m(\mu) \leq \frac{\delta}{2}$, which in turn implies that $s_1(\mu) \leq \frac{\delta}{2}$, $s_2(\mu) \leq \frac{\delta}{2}, \ldots, s_m(\mu) \leq \frac{\delta}{2}$. Then, using Lemma 4.3 once again, we conclude that the neighborhood $\mathcal{N}(\beta, \mu_0)$ stays inside the region R^L_δ for $\beta = \frac{\delta}{4}$. For the central path (2) with $w = e$, the duality gap $c^T x(\mu) - b^T y(\mu) = n\mu$. It is well known (see, e.g., [14]) that $b^T y(\mu)$ is monotonically increasing and $c^T x(\mu)$ is monotonically decreasing along the central path. In our case, $b^T y(\mu) = -y_m(\mu)$ is increasing to 0 and $c^T x(\mu)$ is monotonically decreasing to 0, i.e., $c^T x(\mu) > 0$ for all $\mu > 0$. Then $n\mu = c^T x(\mu) - b^T y(\mu) > y_m$ implies that $\mu > \frac{y_m}{n}$ for any point on the central path. Hence for the central path point for which $y_m(\mu) = \frac{\rho(m)^{m-1}\delta}{2}$, it follows that $\mu_0 > \frac{\rho(m)^{m-1}\delta}{2n}$. Then using the fact that $n = \mathcal{O}(m2^{2m})$, we have $\log(\frac{\mu_1}{\mu_0}) = \mathcal{O}(m)$. The following corollary summarizes our findings.

Corollary 4.1 *Let the neighborhood parameters be given as $\beta_0 < \beta_1 = \frac{1}{16(m+1)}$ for the MTY P-C algorithm. Then there exists a $\mathcal{KM}(m, \rho(m))$ with $n = \mathcal{O}(m2^{2m})$ for which MTY P-C algorithm requires at least $\Omega(2^{m-1})$ predictor steps to reduce the barrier parameter from μ_1 to μ_0 where $\log(\frac{\mu_1}{\mu_0}) = \mathcal{O}(m)$.*

Proof Let $\delta := \frac{1}{4(m+1)}$ and $\beta_1 = \frac{\delta}{4} = \frac{1}{16(m+1)}$. We know from Lemma 4.2 that there exists a $\mathcal{KM}(m, \rho(m))$ with $n = \mathcal{O}(m2^{2m})$ such that $\mathcal{N}(\beta) \subset \mathcal{T}_\delta(m)$. Lemma 4.3 shows that there is an interval $[\mu_0, \mu_1]$ such that the neighborhoods $\mathcal{N}(\beta, \mu_1) \subset R_\delta^U$ and $\mathcal{N}(\beta, \mu_0) \subset R_\delta^L$. Hence starting from an iterate (x^1, y^1, s^1) and μ_1 such that $(x^1, y^1, s^1) \in \mathcal{N}(\beta, \mu_1) \subset R_\delta^U$, in order to reach an iterate (x^0, y^0, s^0) and μ_0 such that $(x^0, y^0, s^0) \in \mathcal{N}(\beta, \mu_0) \subset R_\delta^L$; Propositions 3.2 and 4.2 imply that one needs $\Omega(2^{m-1})$ steps. Since the number of corrector steps is constant, it follows that the number of predictor steps is $\Omega(2^{m-1})$. Moreover the discussion after Lemma 4.3 proves that, we can choose the interval $[\mu_0, \mu_1]$ so that $\log(\frac{\mu_1}{\mu_0}) = \mathcal{O}(m)$. This completes the proof. □

5 A Worst-Case Iteration-Complexity Lower Bound for the Sonnevend's Curvature

In Sect. 4, we proved that the MTY P-C algorithm requires $\Omega(2^{m-1})$ iterations using the larger neighborhood $\mathcal{N}(\beta_1)$ with $\beta_1 = \Omega(\frac{1}{m+1})$. Our goal, in this section, is to derive a lower bound for the Sonnevend's curvature using the tools from the previous section. To this end, we need to examine the constants in Theorem 2.1 more closely.

Lemma 5.1 *Let β_1 be the large neighborhood constant so that $\beta_1 \le \dfrac{1}{400}$ and N be the number of iterations of the MTY P-C algorithm to reduce the barrier parameter from μ_1 to μ_0. Then*

$$N \le \frac{4\sqrt{2}}{\sqrt{\beta_1}} \int_{\mu_0}^{\mu_1} \frac{\kappa(\mu)}{\mu} d\mu + \frac{1}{2\log(1+\frac{\sqrt{\beta_1}}{4})} \log\left(\frac{\mu_1}{\mu_0}\right). \tag{12}$$

Proof See Theorem 2.4 and its proof in [16]. □

The next theorem shows that on the interval $[\mu_0, \mu_1]$, the total Sonnevend's curvature is in comparable order to the number of sharp turns of the central path.

Theorem 5.1 *There is an integer $m_0 > 0$ such that for any $m \geq m_0$, there exists a $\mathcal{KM}(m, \rho(m))$ and interval $[\mu_0, \mu_1]$ such that the Sonnevend's curvature satisfies*

$$\int_{\mu_0}^{\mu_1} \frac{\kappa(\mu)}{\mu} d\mu = \Omega\left(\frac{\sqrt{n}}{(\log n)^2} \log\left(\frac{\mu_1}{\mu_0}\right)\right).$$

Proof Let $\beta_1 = \frac{1}{16(m+1)}$ and choose the parameters of $\mathcal{KM}(m, \rho(m))$ as $\rho(m) = \frac{m}{2(m+1)}$ and $\delta = \frac{1}{8(m+1)}$. Then by Corollary 4.1, we can assume $n = \mathcal{O}(m2^{2m})$. For simplicity we can assume $n = \Theta(m2^{2m})$ by adding some further redundant constraints, if necessary. Then we have $m = \Theta(\log n)$. Corollary 4.1 also makes sure that $\log\left(\frac{\mu_1}{\mu_0}\right) = \Theta(m)$. Since we can extend the interval $[\mu_0, \mu_1]$ so that it still includes all the sharp turns, we will assume that $\log\left(\frac{\mu_1}{\mu_0}\right) = \Theta(m)$. Since $m = \Theta(\log n)$, we have $\log\left(\frac{\mu_1}{\mu_0}\right) = \Theta(\log n)$. Then Corollary 4.1 applies and we have the lower bound for $N \geq 2^{m-1}$. From the facts that $n = \Theta(m2^{2m})$ and $m = \Theta(\log n)$, it follows that $N \geq 2^{m-1} = \Omega(\frac{\sqrt{n}}{\sqrt{\log n}})$.

Now using the bound $\log(1 + \omega) \geq (\log 2)\omega$ for $0 \leq \omega \leq 1$, and $\beta_1 = \frac{1}{16(m+1)}$, we have

$$\frac{1}{2\log(1 + \frac{\sqrt{\beta_1}}{4})} \leq \frac{8\sqrt{m+1}}{\log 2}. \tag{13}$$

If divide both sides of (12) by $\log\left(\frac{\mu_1}{\mu_0}\right)$, we have

$$\frac{N}{\log\left(\frac{\mu_1}{\mu_0}\right)} \leq \frac{\frac{4\sqrt{2}}{\sqrt{\beta_1}} \int_{\mu_0}^{\mu_1} \frac{\kappa(\mu)}{\mu} d\mu}{\log\left(\frac{\mu_1}{\mu_0}\right)} + \frac{1}{2\log(1 + \frac{\sqrt{\beta_1}}{4})}. \tag{14}$$

Then using (13) and (14) together,

$$\frac{N}{\log\left(\frac{\mu_1}{\mu_0}\right)} \leq \frac{\frac{4\sqrt{2}}{\sqrt{\beta_1}} \int_{\mu_0}^{\mu_1} \frac{\kappa(\mu)}{\mu} d\mu}{\log\left(\frac{\mu_1}{\mu_0}\right)} + \frac{8\sqrt{m+1}}{\log 2}. \tag{15}$$

By substituting $\beta_1 = \frac{1}{16(m+1)}$ in (15), we obtain

$$\frac{N}{\log\left(\frac{\mu_1}{\mu_0}\right)} \leq \frac{\leq 16\sqrt{2}\sqrt{m+1} \int_{\mu_0}^{\mu_1} \frac{\kappa(\mu)}{\mu} d\mu}{\log\left(\frac{\mu_1}{\mu_0}\right)} + \frac{8\sqrt{m+1}}{\log 2}. \tag{16}$$

Dividing both sides of (16) by $\sqrt{m+1}$,

$$\frac{N}{\log\left(\frac{\mu_1}{\mu_0}\right)\sqrt{m+1}} \le \frac{\le 16\sqrt{2}\int_{\mu_0}^{\mu_1}\frac{\kappa(\mu)}{\mu}d\mu}{\log\left(\frac{\mu_1}{\mu_0}\right)} + \frac{8}{\log 2}. \tag{17}$$

Finally from the facts that $\log\left(\frac{\mu_1}{\mu_0}\right) = \Theta(m)$, $m = \Theta(\log n)$ and $N = \Omega(\frac{\sqrt{n}}{\sqrt{\log n}})$, we obtain

$$\frac{N}{\log\left(\frac{\mu_1}{\mu_0}\right)\sqrt{m+1}} = \Omega\left(\frac{\sqrt{n}}{(\log n)^2}\right). \tag{18}$$

Hence (18) implies that

$$\int_{\mu_0}^{\mu_1}\frac{\kappa(\mu)}{\mu}d\mu = \Omega\left(\frac{\sqrt{n}}{(\log n)^2}\log\left(\frac{\mu_1}{\mu_0}\right)\right).$$

The proof is complete. □

Corollary 5.1 *For any $\varepsilon > 0$, there is an integer $m_0 > 0$ such that for any $m \ge m_0$, there exists a $\mathcal{KM}(m, \rho(m))$ and interval $[\mu_0, \mu_1]$ such that $\int_{\mu_0}^{\mu_1}\frac{\kappa(\mu)}{\mu}d\mu \ge n^{(\frac{1}{2}-\varepsilon)}\log\left(\frac{\mu_1}{\mu_0}\right)$, where $\log\left(\frac{\mu_1}{\mu_0}\right) = \mathcal{O}(m)$.*

Proof The claim follows from Theorem 5.1 for large m. □

Remark 5.1 Corollary 5.1 yields a negative answer to the question raised by [19], i.e., whether there exists an $\alpha_1 < \frac{1}{2}$ with $\log\left(\frac{\mu_1}{\mu_0}\right) = \Omega(1)$ such that $\int_{\mu_0}^{\mu_1}\frac{\kappa(\mu)}{\mu}d\mu \le n^{\alpha_1}\log\left(\frac{\mu_1}{\mu_0}\right)$ for the class of LO problems.

6 An Iteration-Complexity Lower Bound for MTY P-C Algorithm with Constant Neighborhood Opening

In practice, the MTY P-C algorithm operates in a larger neighborhood where β_1 is a constant. In order to conclude an iteration-complexity lower bound for MTY P-C algorithm with constant neighborhood opening β_1 by using Theorem 2.1, we need to show that there is a constant $\nu > 0$ with $\kappa(\mu) \ge \nu$ for $\mu \in [\mu_0, \mu_1]$ for $\mathcal{KM}(m, \rho(m))$. While this appears to hold numerically, proving it is much more difficult. To go around this difficulty, we exploit a trick introduced by [15]. The idea is to use one-dimensional LO problems, where it is easier to calculate the central path and its corresponding $\kappa(\mu)$; and to use LO problems with scaled objectives with block diagonal constraints. For the details, we refer the reader to Appendix section.

Recall that by Corollary 5.1, we know there exists a $\mathcal{KM}(m, \rho(m))$ and an interval $[\mu_0, \mu_1]$ such that $\int_{\mu_0}^{\mu_1} \frac{\kappa(\mu)}{\mu} d\mu \geq n^{\left(\frac{1}{2}-\varepsilon\right)} \log\left(\frac{\mu_1}{\mu_0}\right)$. Here $n = \mathcal{O}(m2^{2m})$ and $\frac{\mu_1}{\mu_0} = \mathcal{O}(\log n)$. Now by using Lemma 8.2 and Proposition 8.1, we can embed $\mathcal{KM}(m, \rho(m))$ in a block diagonal LO problem at the expense of increasing the size of the problem by at most $\overline{n} := n + \mathcal{O}(m + \log m)$. Denote by $\overline{\mathcal{KM}}(\overline{m})$ this hybrid construction with $\mathcal{KM}(m, \rho(m))$ embedded in. Since $\overline{n} = \mathcal{O}(n)$, we have the following:

Theorem 6.1 *For any $\varepsilon > 0$, there exists a positive integer m_0 such that for any $\overline{m} \geq m_0$, there exists an LO problem $\overline{\mathcal{KM}}(\overline{m})$ and an interval $[\mu_0, \mu_1]$ with the following properties:*

- $\frac{\mu_1}{\mu_0} = \mathcal{O}\left(m2^{2m}\right)$.
- *Let $\beta_0 < \beta_1 \leq \frac{1}{400}$ be the constant neighborhood $\mathcal{N}(\beta)$ parameters. Then, the MTY P-C algorithm on this neighborhood requires $\Omega\left(n^{\left(\frac{1}{2}-\varepsilon\right)} \log\left(\frac{\mu_1}{\mu_0}\right)\right)$ predictor steps.*

Proof Consider the $\mathcal{KM}(m, \rho(m))$ cube from Corollary 5.1. Then by using Lemma 8.2 and Proposition 8.1, we can embed $\mathcal{KM}(m, \rho(m))$ in a block diagonal LO problem with size $\overline{n} := n + \mathcal{O}(m + \log m)$ and $\overline{m} = \mathcal{O}(m)$. Note that since the interval $[\mu_0, \mu_1]$ comes from $\mathcal{KM}(m, \rho(m))$, the first claim in the theorem follows from Corollary 5.1. Also, since for $\overline{\mathcal{KM}}(\overline{m})$, there exists a constant $\nu > 0$ for all $\mu \in [\mu_0, \mu_1]$ with the corresponding $\overline{\kappa}(\mu) \geq \nu$, Theorem 2.1 implies the first claim. This completes the proof. □

7 Conclusion and Future Work

It is an open question whether there is an interior-point algorithm for LO problems with $\mathcal{O}(n^{\alpha_1} \log(\frac{\mu_1}{\mu_0}))$ iteration-complexity upper bound for $\alpha_1 < \frac{1}{2}$ to reduce the barrier parameter from μ_1 to μ_0. In this regard, a related open question raised by Stoer et al. [16] was whether there is an $\alpha_1 < \frac{1}{2}$ with $\int_{\mu_0}^{\mu_1} \frac{\kappa(\mu)}{\mu} d\mu \leq n^{\alpha_1} \log\left(\frac{\mu_1}{\mu_0}\right)$ for all LO problems. This paper provides a negative answer to the latter question. We also show that for the MTY P-C algorithm, the classical iteration-complexity upper bound is tight. Future work would be to investigate whether an analogous result could be derived to the case of long step IPMs.

In this paper, we establish that for the central path of the carefully constructed redundant Klee-Minty cubes, both the geometric curvature and the Sonnevend's curvature of the central path are essentially in the order of $\Omega(\sqrt{n})$. In a recent work,

Mut and Terlaky [9] show the existence of another class of LO problems where a large geometric curvature of the central path implies a large Sonnevend's curvature. These two important cases suggest that it might be possible to prove this implication in a more general setting.

Acknowledgements Research supported by a Start-up grant of Lehigh University. It is also supported by TAMOP-4.2.2.A-11/1KONV-2012-0012: Basic research for the development of hybrid and electric vehicles. The TAMOP Project is supported by the European Union and co-financed by the European Regional Development Fund.

Appendix

Lemma 8.1 *For large enough r, there is one-dimensional LO problem with $(r+1)$ constraints for which $\tau_1\sqrt{r} \le \kappa(\mu) \le \tau_2\sqrt{r}$ for any $\mu \in [\alpha_1, \alpha_2]$, where $\alpha_1 = \frac{1}{r-\frac{\sqrt{r}}{4}}$ and $\alpha_2 = \frac{1}{r-\sqrt{r}}$ for some constants $\tau_1, \tau_2 \ge 0$.*

Proof Consider the problem $\min\{\, y : \; y \le 1 \text{ and, } \; y \ge 0 \text{ counted } r \text{ times}\}$. The construction is given in [15], p:551. Consider the interval $[\alpha_1, \alpha_2]$, where $\alpha_1 = \frac{1}{r-\frac{\sqrt{r}}{4}}$ and $\alpha_2 = \frac{1}{r-\sqrt{r}}$. Let $s_0(\mu) = 1 - y(\mu)$. Then it is shown in [15], p. 551 that, $\dfrac{\dot{s}_0(\mu)}{s_0(\mu)} \ge \dfrac{r^2}{3\sqrt{r}}$ on $[\alpha_1, \alpha_2]$. This implies $\dfrac{\mu\dot{s}_0(\mu)}{s_0(\mu)} = \Omega(\sqrt{r})$ on $[\alpha_1, \alpha_2]$. Then, from Proposition 2.1 part *1.*, we have $\kappa(\mu) = \Omega(\sqrt{r})$ for all $\mu \in [\alpha_1, \alpha_2]$. The proof is complete. □

Proposition 8.1 *Consider the LO problems*

$$\begin{array}{lll} \min\ (c^1)^T x & & \min\ (c^2)^T x \\ \text{s.t.} \quad A^1 x^1 = b^1 & \textit{and} & \text{s.t.} \quad A^2 x^2 = b^2 \\ \qquad\quad x^1 \ge 0, & & \qquad\quad x^2 \ge 0, \end{array} \tag{19}$$

with the corresponding $\kappa^1(\mu)$ and $\kappa^2(\mu)$ on the interval $[\mu_0, \mu_1]$. Then for the problem

$$\begin{array}{l} \min\ c^T x \\ \text{s.t.} \quad Ax = b \\ \qquad\quad x \ge 0, \end{array} \tag{20}$$

with the corresponding $\overline{\kappa}(\mu)$ where $c = \begin{bmatrix} c^1 \\ c^2 \end{bmatrix}$, $b = \begin{bmatrix} b^1 \\ b^2 \end{bmatrix}$ and $A = \begin{bmatrix} A^1 & 0 \\ 0 & A^2 \end{bmatrix}$, on $[\mu_0, \mu_1]$, we have $\overline{\kappa}(\mu) \ge \kappa^i(\mu)$ for $i = 1, 2$.

Proof Let $\left(x^1(\mu), y^1(\mu), s^1(\mu)\right)$ and $\left(x^2(\mu), y^2(\mu), s^2(\mu)\right)$ be the central paths in (19). Then the term $\overline{\kappa}(\mu)$ for the combined problem (20) becomes $\overline{\kappa}(\mu) = \left|\left|[\mu\dot{x}^1\dot{s}^1, \mu\dot{x}^2\dot{s}^2]\right|\right|^{\frac{1}{2}} \ge \kappa^i(\mu)$ for $i = 1, 2$. □

Proposition 8.2 *Let $\eta > 0$ and consider the central path ((2) and its $\kappa(\mu)$. Let $(\hat{A}, \hat{b}, \hat{c})$ be another problem instance, where $(\hat{A}, \hat{b}, \hat{c}) = (A, \frac{b}{\eta}, c)$ with its corresponding $\hat{\kappa}(\mu)$. Then, we have*

$$\hat{\kappa}(\mu) = \kappa(\eta\mu), \ \mu \in \left[\frac{\mu_0}{\eta}, \frac{\mu_1}{\eta}\right]. \tag{21}$$

Proof Using (2), it is straightforward to verify that the central path $(\hat{x}(\mu), \hat{y}(\mu), \hat{s}(\mu))$ of the new problem satisfies $\hat{x}(\mu) = \dfrac{x(\eta\mu)}{\eta}$, $\hat{y}(\mu) = y(\eta\mu)$ and $\hat{s}(\mu) = s(\eta\mu)$. Using the definition of $\kappa(\mu)$, we get $\hat{\kappa}(\mu) = \kappa(\eta\mu)$. Hence the claim follows. □

Lemma 8.2 *Given an interval $[\mu_0, \mu_1]$ and a constant $\nu > 0$, there exists an LO problem of size $n = \Theta\left(\log(\frac{\mu_1}{\mu_0})\right)$ such that $\overline{\kappa}(\mu) \geq \nu$ for all $\mu \in [\mu_0, \mu_1]$. The hidden constant in $n = \Theta\left(\log(\frac{\mu_1}{\mu_0})\right)$ depends on ν.*

Proof Let a constant $\nu > 0$ and an interval $[\mu_0, \mu_1]$ be given. For the given $\nu > 0$, by Lemma 8.1, there exists an LO problem with its $\kappa(\mu) \geq \nu$ on an interval $\mu \in [\alpha_1, \alpha_2]$. By applying Proposition 8.2 for $\eta := \frac{\alpha_1}{\left(\frac{\alpha_2}{\alpha_1}\right)^i \mu_0}$ for $i = 0, 1, \ldots, k$, we find $(k-1)$ scaled LO problems with their corresponding $\kappa^i(\mu)$, $i = 0, 1, \ldots, k-1$ such that $\kappa^i(\mu) = \kappa(\eta\mu)$ on $\mu \in \left[(\frac{\alpha_2}{\alpha_1})^i \mu_0, (\frac{\alpha_2}{\alpha_1})^{i+1} \mu_0\right]$, for $i = 0, 1, \ldots, k-1$. Then by using Proposition 8.1, we can obtain a block diagonal LO problem with its $\overline{\kappa}(\mu) \geq \kappa^i(\mu) \geq \nu$ for $i = 0, 1, \ldots, k-1$ for any $\mu \in \left[\mu_0, \left(\frac{\alpha_2}{\alpha_1}\right)^k \mu_0\right]$. In order to have $\overline{\kappa}(\mu) \geq \nu$ for any $\mu \in [\mu_0, \mu_1]$, it is then enough to have $\left(\frac{\alpha_2}{\alpha_1}\right)^k \mu_0 \geq \mu_1$. This is true if and only if $k \log\left(\frac{\alpha_2}{\alpha_1}\right) \geq \log\left(\frac{\mu_1}{\mu_0}\right)$. Since by Lemma 8.1, the ratio $\frac{\alpha_2}{\alpha_1}$ is a constant depending only on the given ν, the number of blocks k needed is $\Theta\left(\log(\frac{\alpha_2}{\alpha_1})\right)$. Also since the size of the LO problem with its $\kappa(\mu)$ is a constant only determined by ν, the size of the problem is $n = \Theta\left(\log(\frac{\mu_1}{\mu_0})\right)$ to achieve $\overline{\kappa}(\mu) \geq \nu$ for all $\mu \in [\mu_0, \mu_1]$. This completes the proof. □

References

1. Deza, A., Nematollahi, E., Peyghami, R., Terlaky, T.: The central path visits all the vertices of the Klee-Minty cube. Optim. Methods Softw. **21**(5), 851–865 (2006)
2. Deza, A., Nematollahi, E., Terlaky, T.: How good are interior point methods? Klee-Minty cubes tighten iteration-complexity bounds. Math. Program. **113**(1), 1–14 (2008)
3. Huhn, P., Borgwardt, K.H.: Interior-point methods: worst case and average case analysis of a phase-i algorithm and a termination procedure. J. Complex. **18**(3), 833–910 (2002)
4. Jansen, B., Roos, C., Terlaky, T.: A short survey on ten years interior point methods. Technical report 95–45, Delft University of Technology, Delft, The Netherlands (1995)

5. Karmarkar, N.: A new polynomial-time algorithm for linear programming. Combinatorica **4**(4), 373–395 (1984)
6. Megiddo, N., Shub, M.: Boundary behavior of interior point algorithms in linear programming. Math. Oper. Res. **14**(1), 97–146 (1989)
7. Mizuno, S., Todd, M., Ye, Y.: Anticipated behavior of path-following algorithms for linear programming. Technical report 878, School of Operations Research and Industrial Engineering, Ithaca, New York (1989)
8. Mizuno, S., Todd, M.J., Ye, Y.: On adaptive-step primal-dual interior-point algorithms for linear programming. Math. Oper. Res. **18**(4), 964–981 (1993)
9. Mut, M., Terlaky, T.: An analogue of the Klee-Walkup result for Sonnevend's curvature of the central path. J. Optim. Theory Appl. **169**(1), 17–31 (2016)
10. Nematollahi, E., Terlaky, T.: A redundant Klee-Minty construction with all the redundant constraints touching the feasible region. Oper. Res. Lett. **36**(4), 414–418 (2008)
11. Nematollahi, E., Terlaky, T.: A simpler and tighter redundant Klee-Minty construction. Optim. Lett. **2**(3), 403–414 (2008)
12. Nesterov, Y., Nemirovskii, A.: Interior-Point Polynomial Algorithms in Convex Programming, vol. 13. SIAM, Philadelphia (1994)
13. Potra, F.A.: A quadratically convergent predictor-corrector method for solving linear programs from infeasible starting points. Math. Program. **67**(1–3), 383–406 (1994)
14. Roos, C., Terlaky, T., Vial, J.P.: Interior Point Methods for Linear Optimization. Springer, New York (2006)
15. Sonnevend, G., Stoer, J., Zhao, G.: On the complexity of following the central path of linear programs by linear extrapolation II. Math. Program. **52**, 527–553 (1991)
16. Stoer, J., Zhao, G.: Estimating the complexity of a class of path-following methods for solving linear programs by curvature integrals. Appl. Math. Optim. **27**, 85–103 (1993)
17. Todd, M.J.: A lower bound on the number of iterations of primal-dual interior-point methods for linear programming. Technical report, Cornell University Operations Research and Industrial Engineering (1993)
18. Todd, M.J., Ye, Y.: A lower bound on the number of iterations of long-step primal-dual linear programming algorithms. Ann. Oper. Res. **62**(1), 233–252 (1996)
19. Zhao, G.: On the relationship between the curvature integral and the complexity of path-following methods in linear programming. SIAM J. Optim. **6**(1), 57–73 (1996)

Optimization of Inventory and Distribution for Hip and Knee Joint Replacements via Multistage Stochastic Programming

Mohammad Pirhooshyaran and Lawrence V. Snyder

Abstract We introduce a multistage stochastic programming model to optimize the distribution–production network of medical devices; in particular, artificial hip and knee joints for orthopedic surgery. These devices are distributed to hospitals in kits that contain multiple sizes of the joint; the surgeon uses one device from the kit and then returns the rest of the kit to the distributor, which replaces the part that has been removed and distributes the kit anew. Therefore, the distribution problem for artificial joints has a shareability property and thus is related to closed-loop supply chains. We assume that demands for the devices follow a discrete probability distribution and therefore we use scenarios to model the random demands over time. We compare the results of our optimization model to an approximation of the simple distribution strategy that our industry partner currently uses. The proposed approach outperforms the present approach in terms of optimal cost. We also explore the sensitivity of the model's computation time as the numbers of scenarios, hospitals, and time periods change. Finally, we extend the model to investigate the production of shareable items in sharing systems using a numerical example.

Keywords Hip and knee joint replacement logistics · Sharing systems Production–distribution planning · Multistage stochastic programming Healthcare systems

M. Pirhooshyaran (✉) · L. V. Snyder
Mohler Lab, Department of Industrial and Systems Engineering, Lehigh University, 200 West Packer Ave., Bethlehem, PA 18015, USA
e-mail: mop216@lehigh.edu

L. V. Snyder
e-mail: lvs2@lehigh.edu

J. D. Pintér and T. Terlaky (eds.), *Modeling and Optimization: Theory and Applications*, Springer Proceedings in Mathematics & Statistics 279,
https://doi.org/10.1007/978-3-030-12119-8_7

1 Introduction

The last two decades have witnessed a dramatic increase in the number of hip and knee replacement operations around the world. Both are currently among the most common orthopedic surgeries. Aimed at improving the functionality and quality of life of patients, more than half a million knee and hip replacements are performed in the United States annually. Considering that population aging and obesity are both on the rise, the requests for joint replacement operations are predicted to exceed two million by the end of 2025 [19, 45]. On the other hand, the average cost of a total joint replacement is up to $40,000 in the United States and around $12,000 in the majority of European countries. The total cost of the components to be used in a joint replacement surgery can be up to $5,000 [31, 40].

It is common in the orthopedic industry for device manufacturers, or third-party distribution companies, to own and maintain the inventory of the devices. Typically, most of the inventory is either located at a hospital for use in a surgery in the near future, or in transit to or from the distributor's warehouse or from one hospital to another; very little inventory is stored for any significant amount of time at any location since the parts are so expensive.

Each surgery uses several different device components, and the size needed for each component in order to fit the patient's body is not precisely known prior to the surgery. Doctors use gender, age, body weight, and so on, as well as MRI and other imaging, to estimate the size beforehand, but there is still inherent uncertainty [2, 5, 31, 40]. Therefore, the devices are distributed in kits, with each kit containing multiple sizes of each component. The surgeon removes the components needed and uses them during the surgery. The kit, minus one unit of each component, is then returned to the distributor, which replenishes the missing components and distributes the kit anew. Thus, the supply chain for orthopedic devices has many similarities to supply chains for reusable or shareable goods, i.e., to closed-loop or reverse logistics systems. These systems pertain to items that are environmentally detrimental, technologically advanced, or demanded for a very short period of time [8, 15].

Most of the work done in supply chain design related to reusing products is limited to reprocessing and refurbishing of items that are out of order while there is still value in them. These approaches constitute reverse logistics in closed-loop supply networks. This chapter, however, considers the redistribution of orthopedic device kits that are in perfect order as long as there is a demand for them. Therefore, a conceptual difference exists between reusability for the purpose of this chapter and the concept of reusing the remaining value of an item at its end of life (EOL).

In this chapter, we propose a novel multistage stochastic linear programming model for distributing items within a reusable-item supply chain. Our model is designed for orthopedic devices but could be adapted for use in other types of supply chains. The objective here is to develop an optimization framework that explicitly accounts for the unique aspects of hip and knee joint operations while also tackling the inherent uncertainties from the point of view of the production–distribution system. We consider two approaches for solving the model. In the first, we assume the number of available items is fixed and the aim is to optimize the distribution only, using multistage stochastic programming. In the second, we extend the model to include the decision of how many devices to produce.

The remainder of this chapter is organized as follows: Sect. 2 summarizes the literature review. Section 3 gives a detailed description of the problem and how we model it. Section 4 presents the numerical study and our approaches for solving the model. Section 5 expands the model to consider the production decision. Finally, Sect. 6 contains our discussion and conclusions.

2 Literature Review

In this section, we discuss the two most relevant categories of literature, on the production–distribution of reusable products and on modeling uncertainty.

2.1 Production–Distribution of Reusable Products

The conventional definition of a supply chain, which involves transferring goods or services from producers to consumers, is no longer valid for products that must be sent back from their final destination to other parts of the supply chain. Such activities are called reverse logistics. To make supply chains more sustainable, manufacturers often consider several possible closed loops in their supply chain design and optimize the way that products are distributed in these loops [21, 38]. Moreover, to mitigate the adverse environmental effects of supply chains, government regulations and societal pressure have led manufacturers to design their supply chain networks in order to manage both forward and reverse item flows [23].

In addition, sharing systems such as bicycle and vehicle sharing have grown in popularity in recent years. The task of transporting reusable products within the network to restore inventory levels at distribution centers to a satisfactory threshold is called inventory rebalancing [14]. Therefore, distribution centers, which serve as links between manufacturers and customers, play an important role in rebalancing the inventory of sharing systems [37].

Several modeling approaches have been proposed for rebalancing inventory of reusable products in sharing supply chains. These include Markov Chain formulations [29, 37], decomposition approaches [7], game-theoretic approaches [12], and mathematical optimization [30, 49]. Most of these shared-mobility models optimize the distribution of available shareable products while leaving open the important question of the optimal production quantities. In other words, sharing systems are usually studied as extensions of vehicle routing problems rather than production problems.

In closed-loop supply chains, on the other hand, most studies focus on production–distribution networks, with the aim of extracting any remaining value out of partially used products. These loops may be designed to connect refurbishing centers and manufacturers [22, 47] to retrieve the remaining raw materials. The loops may be designed to connect disposal centers as a back end of a network to dispose of unusable parts [34, 44].

2.2 Modeling Uncertainty

There are different approaches for modeling uncertainty in supply chain optimization problems. These include Markov decision processes (MDP), e.g., [28, 32], robust optimization (RO) [4], and multistage stochastic programming (MSP) [16] and fuzzy approaches [20]. See [41, 42] for reviews of these and other approaches as applied to facility location and network design problems. All of these approaches are generalizations of deterministic mathematical programming models in which the actual values of some model parameters are not known at the time decisions are made. Such models often include exponentially many decision variables with numerous potential values, discrete time intervals for making decisions, implementation of expectation or variance functionals into objectives, and known (or partially known) distributions which uncertain parameters follow [6]. A solution is "acceptable" if it performs acceptably under any possible realization of the unknown parameters. The definition of acceptable performance differs from application to application [41]. If probability information is completely known, uncertainty is limited to the use of (continuous or discrete) probability distributions on the parameters, which is when MSP or MDP is used most often [43]. Otherwise, if no probability information is given, then the unknown parameters are typically confined to lie within some intervals, in which case RO is often used [3]. The case where we have partial information of the parameters, such as the moments of the distribution, lies in a category called distributionally robust optimization [48].

In this study, volatility in demand can have a noticeable impact on the supply network, especially when we are dealing with limited supply. The exact number of operations at each hospital is not a deterministic quantity for the distributor to consider. Not only do hospitals face emergency operations due to injuries such as a broken knee, as well as sudden cancelations by the patient, but even with a regular schedule, hospitals typically do not request the devices they need from the distributor too far in advance. Therefore, at the beginning of the decision horizon, the distributor must make decisions at best based on historical data and/or expert opinions about how to distribute the available products among the hospitals. Their aim is to minimize the cost as well as to maintain a sufficient level of patient satisfaction under uncertainty.

Here, we model the uncertain demand as a random variable with a known, discrete probability distribution, which we represent using a scenario tree. In particular, a "scenario" represents a specific realization of the future, i.e., a specification of the value of each random parameter in every period of the time horizon. If the distribu-

Table 1 Different approaches for reusable products

Articles	Network structure	Optimal production	Rebalancing inventory	Modeling	Demand
Devika et al. [17]	Closed-loop	Considered	×	MILP[a]	Deterministic
Anvari and Turkay [1]	Closed-loop	×	×	MO-MILP[b]	Deterministic
Rezaee et al. [36]	Forward	Considered	×	TSLS[c]	Stochastic
Haddadsisakht et al. [23]	Closed-loop	Considered	×	TSHRS[d]	Stochastic-Robust
Schuijbroek et al. [37]	Closed-loop	×	Considered	Heuristic	Stochastic
Chiariotti et al. [11]	Closed-loop	×	Considered	BDP[e]	Deterministic
Raviv et al. [35]	Closed-loop	×	Considered	CMC[f]	Stochastic
Chemla et al. [9]	Closed-loop	×	Considered	ILP	Stochastic
Spiliotopoulou et al. [46]	Forward	Considered	Considered	Game theory	Deterministic
Our proposed method	Closed-loop	Considered	Considered	MSLP[g]	Stochastic

[a]Mixed-integer linear programming
[b]Multi-objective MILP
[c]Two-stage linear stochastic
[d]Three-stage hybrid robust-stochastic
[e]Birth–death processes
[f]Continuous-time Markov chain
[g]Multistage stochastic linear programming

tion support is infinite, an accurate construction of the scenario tree would obviously require an infinite number of nodes. Otherwise, enumeration of finitely many outcomes allows us to express all scenarios in a tree, with each scenario consisting of a path from the root node of the tree to one of its end nodes (leaves) [10, 16, 25].

Table 1 summarizes the recent literature on sharing supply chains, indicating the approach that each paper takes regarding the supply network structure, optimal production, rebalancing inventory, modeling approach, and demand uncertainty. The features of our proposed method are included in the last line of the table.

As seen in Table 1, no work has been published on closed-loop supply networks of sharing systems that considers optimal production as well as distributor–customer loops in order to rebalance the inventory prior to our proposed method.

3 Problem Definition and Modeling

As discussed above, before a given surgery there is uncertainty about the sizes of the device components that will be needed. Therefore, distributors send kits of devices containing a variety of shapes and sizes in order to protect against the uncertainty and prevent stockouts, which are extremely undesirable in healthcare settings. Every time a hospital performs a given type of implant surgery, the surgeon removes the devices needed and releases the kit to be reused. The kit can be transported back to the distributor, and from there it is sent to another hospital. Alternately, if the first hospital has another surgery scheduled in the near future, the kit can remain at that hospital in order to save transportation costs. In either case, the devices that have been removed are replaced by the distributor before the next surgery. In this sense, we can model the devices as shareable products. In our research, we ignore the replenishment of the used components back into the kit, and instead model only the movement of kits throughout the systems.

Our initial model focuses on the distribution centers (DCs) and hospitals, as well as the inventory-rebalancing loops among them, assuming the number of kits is predefined. Figure 1 depicts the main loops, between DCs and hospitals. These loops constitute a connected graph among all possible nodes. Once a kit is used by a hospital, it can be kept at the hospital for another surgery, transported directly to another hospital, or transported to another hospital via the DC (if there is no direct path between the two hospitals). If there is no upcoming demand for the kit, it should be stored in the DC.

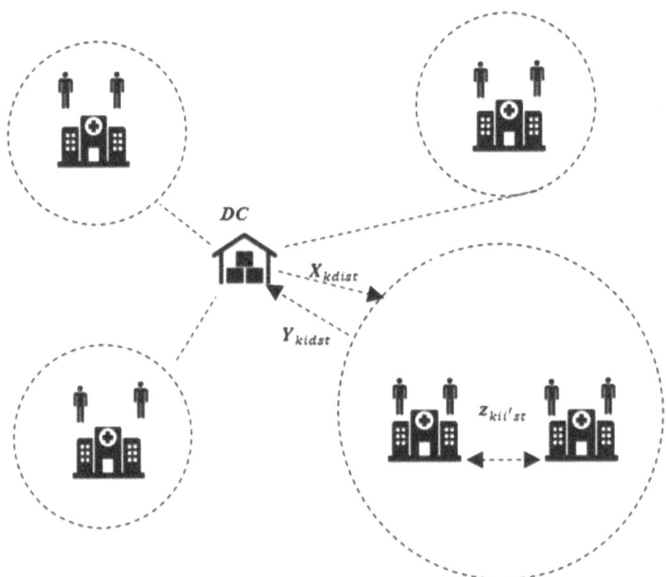

Fig. 1 Distribution center and hospitals (rebalancing loops)

We further assume:

- The capacities of all facilities are fixed and known.
- Transportation costs are fixed and known.
- Demands may be either deterministic or stochastic.
- DC and hospital locations are known a priori.
- Demand shortages incur a penalty cost in the objective function.
- Transporting a kit from one location (hospital or DC) to another takes exactly one time period.

We use the following notation:
Indices:

k = index for type of implant, $k \in \mathbb{K}$
t = index for time periods, $t \in \mathbb{T}$
s = index for scenarios, $s \in \mathbb{S}$
d = index for DCs, $d \in \mathbb{D}$
i,i' = indices for hospitals, $i, i' \in \mathbb{I}$

Variables:

X_{kdist} = quantity of type-k implant kits shipped from DC d to hospital i in scenario s and period t
Y_{kidst} = quantity of type-k implant kits shipped from hospital i to DC d in scenario s and period t
$Z_{kii'st}$ = quantity of type-k implant kits shipped from hospital i to hospital i' in scenario s and period t
Δ_{kist} = unsatisfied demand of type-k implant kits at hospital i in scenario s and period t

Parameters:

b_{kdt} = capacity of type-k implant kits at DC d in period t
b_{kit} = maximum number of type-k implant kits that hospital i can order in period t
c_{kdi} = per-unit transportation cost of type-k implant kits from DC d to hospital i
c_{kid} = per-unit transportation cost of type-k implant kits from hospital i to DC d
$c_{kii'}$ = per-unit transportation cost of type-k implant kits from hospital i to hospital i'
d_{kist} = demand of type-k implant kits at hospital i in scenario s and period t
p_s = probability that scenario s occurs
v_k = number of type-k implant kits available in the system at the beginning of horizon
ζ_{kit} = per-unit penalty cost for unsatisfied demand of type-k implant kits at hospital i in period t

The problem can be formulated as a multistage stochastic optimization problem as follows:

$$\mathbb{Z}(v_k) : \min \sum_s p_s \left(\sum_k \sum_d \sum_i \sum_t c_{kdi} X_{kdist} + \sum_k \sum_i \sum_d \sum_t c_{kid} Y_{kidst} + \sum_k \sum_i \sum_{i'} \sum_t c_{kii'} Z_{kii'st} \right. \quad (1)$$

$$\left. + \sum_k \sum_i \sum_t \zeta_{kit} \Delta_{kist} \right) \quad (2)$$

$$\text{subject to} \sum_i X_{kdist} \leq b_{kdt} \ \forall k \in \mathbb{K},\ d \in \mathbb{D}\ s \in \mathbb{S} \text{ and } t \in \mathbb{T} \quad (3)$$

$$\sum_d Y_{kidst} + \sum_{i'} Z_{kii'st} \leq b_{kit} \ \forall k \in \mathbb{K},\ i \in \mathbb{I},\ s \in \mathbb{S} \text{ and } t \in \mathbb{T} \quad (4)$$

$$\sum_{t^\star=1}^{t} \left(\sum_d X_{kdist^\star} + \sum_{i'} Z_{ki'ist^\star} - \sum_d Y_{kidst^\star} \right) + \Delta_{kist} \geq d_{kist} \quad (5)$$

$$\forall k \in \mathbb{K},\ i \in \mathbb{I}, s \in \mathbb{S} \text{ and } t \in \mathbb{T}$$

$$\sum_{t^\star=1}^{t} \left(\sum_i X_{kdist^\star} - \sum_i Y_{kids(t^\star-1)} \right) \leq v_k \ \forall k \in \mathbb{K},\ i \in \mathbb{I}, s \in \mathbb{S} \text{ and } t \in \mathbb{T} \quad (6)$$

$$X_{kdis_m 1} = X_{kdis_n 1} \ \forall (s_m, s_n) \in \mathbb{S} \times \mathbb{S},\ k \in \mathbb{K},\ d \in \mathbb{D} \text{ and } i \in \mathbb{I} \quad (7)$$

$$X_{kdis_m t} = X_{kdis_n t} \ \forall (s_m, s_n) \in \bar{\Omega} \quad (8)$$

$$k \in \mathbb{K},\ i \in \mathbb{I}, \text{ and } t \in \mathbb{T} \setminus \{1\}$$

$$X_{kidst}, Y_{kdist}, Z_{kii'st} \geq 0, \ \forall k \in \mathbb{K}, d \in \mathbb{D},\ i \in \mathbb{I}\ s \in \mathbb{S} \text{ and } t \in \mathbb{T}. \quad (9)$$

Where

$$\bar{\Omega} = \left\{ (s_m, s_n) \in \mathbb{S} \times \mathbb{S} \mid \left((d_{kis_m 1}, \ldots, d_{kis_m (t-1)}) = (d_{kis_n 1}, \ldots, d_{kis_n (t-1)}) \right) \right\}.$$

The model minimizes the total expected cost of the proposed sharing network with respect to the scenarios. Equation (1) computes the expected transportation cost, while Eq. (2) computes the expected penalty cost of unsatisfied demands. Constraints (3)–(4) enforce the capacities at distribution centers and hospitals. Note that for hospitals, we assume there is an upper bound on the number of items that can be ordered at the same time. Constraints (5) require the demand for all the hospitals to be satisfied, or that unmet demands are included in Δ_{kit}. In these constraints, the indices t and $t^\star$ keep track of the kits as they move into and out of the hospital over the time horizon, in order to calculate the current on-hand inventory at the hospital. Constraints (6) enforce the availability of type-k implant kits. Here, it has been assumed that over the planning horizon, the kits remain in perfect working order and do not need to be replaced or repaired. Constraints (7) and (8) are nonanticipativity constraints, which are standard for multistage stochastic optimization problems; see [13, 18]. Finally, constraints (9) require the variables to be nonnegative. The model can be reduced to a network flow problem, for which it is well known that as long as all of the demand data are integers, there is an optimal solution to the problem consisting of only integers. Therefore, it is sufficient for us to require only nonnegativity.

4 Numerical Studies

In this section, we present computational results from our proposed model on a small instance and compare them to an emulation of the strategy used by our industry partner. We tested the model on instances with between 2 and 10 hospitals and 5 or 7 time periods. We assumed that the demand at each hospital in each period is distributed uniformly in $\{0, \ldots, 5\}$. We generated scenarios by sampling from the demand distribution and assumed that each scenario has equal probability. To obtain a scenario set of manageable size, we use the sample average approximation (SAA) method [39]. That is, assuming the random parameters follow a discrete uniform distribution, we generate an unbiased sample of scenarios from a pool of iid scenario realizations. We assumed that $|\mathbb{K}| = 2$ (one type of hip and one type of knee implant) and that there is a single DC ($|\mathbb{D}| = 1$). We prohibited the shipment of kits from one hospital directly to another (i.e., we forced $Z_{kii'st} = 0$) to model the situation in which kits must be inspected by the distributor before sending them to a new hospital. We set $c_{k1i} = 10i^2$, $b_{k11} = 2.5|\mathbb{I}|$, and $\zeta_{kit} = 3c_{k1i}$, for all $t \in \mathbb{T}$.

We formulated the model in AMPL Version 20070505 and solved it using CPLEX 12.7.0.0. The total number of variables is $3|\mathbb{K}||\mathbb{I}||\mathbb{D}||\mathbb{S}||\mathbb{T}|$, which varies from 3000 up to 4,20,000 in the instances we tested. The number of constraints is not deterministic because the number of nonanticipativity constraints depends directly on the realizations of the demands. However, excluding the nonanticipativity constraints, there are $3|\mathbb{K}||\mathbb{I}||\mathbb{D}||\mathbb{S}||\mathbb{T}| + 3|\mathbb{K}||\mathbb{I}||\mathbb{S}||\mathbb{T}| + |\mathbb{K}||\mathbb{D}||\mathbb{S}||\mathbb{T}|$ constraints, which varies from 6500 up to 8,54,000 in our instances.

Table 2 reports the elapsed CPU time required to solve the model, using the default termination parameters. From the table, it is clear that the model is somewhat sensitive to increases in the number of hospitals and even more sensitive to the number of scenarios. For example, when $|\mathbb{S}| = 50$ or 100, as $|\mathbb{I}|$ changes from 2 to 10, the elapsed time roughly doubles. In contrast, when $|\mathbb{T}| = 5$, as $|\mathbb{S}|$ increases from 50 to 1000 (a factor of 20), the elapsed time increases by a factor of 68 for $|\mathbb{T}| = 5$ and 238 for $|\mathbb{T}| = 7$. This steep increase in elapsed time when $|\mathbb{S}|$ increases is caused in large part by the nonanticipativity constraints, whose cardinality is quadratic in the number of scenarios [24].

Table 2 Total elapsed time using AMPL/CPLEX (seconds), $d_{kist} \sim U\{0, 5\}$

Scenarios	Number of stages							
	$\mathbb{T} = 5$				$\mathbb{T} = 7$			
	Number of hospitals							
	$\mathbb{I} = 2$	$\mathbb{I} = 3$	$\mathbb{I} = 5$	$\mathbb{I} = 10$	$\mathbb{I} = 2$	$\mathbb{I} = 3$	$\mathbb{I} = 5$	$\mathbb{I} = 10$
$\mathbb{S} = 50$	1.91	2.34	3.16	4.85	4.11	4.71	5.52	7.44
$\mathbb{S} = 100$	6.04	7.67	9.87	11.00	10.75	13.07	15.29	20.08
$\mathbb{S} = 500$	17.70	20.06	39.75	75.59	66.39	93.73	154.37	306.04
$\mathbb{S} = 1000$	56.72	147.48	216.50	601.97	558.84	820.31	1314.28	–[a]

[a]This cell is empty as a result of an "out-of-memory" error

Our industry partner at present transports implant kits to hospitals with scheduled hip and knee surgeries and then transports the kits back to the DC after each surgery, regardless of the fact that it might be better to keep the product inside the hospital for future use. This leads to unnecessarily high transportation and penalty costs. We model their approach, which we call the *present approach*, by optimizing the distribution of kits in each time period independently and myopically, without considering future needs. We wish to investigate the impact of stochasticity in the proposed model against present approach.

One drawback of the SAA approach is that the solution returned by the model only provides decision variable values for the scenarios that are actually sampled. If the observed scenarios lie outside of the sampled set (which is extremely likely), the results provide no guidance about what decisions to take. To address this, and to facilitate a head-to-head comparison between the results of our model and the present approach, we propose a *folding horizon* framework. In any time period t, we optimize the decision variables for periods $t,\ t+1, ..., \mathbb{T}$ under uncertain demands as represented by a fixed number $|\mathbb{S}|$ of scenarios. Then, we implement the optimal decision variables in period t and observe a new demand realization. We update the initial conditions, such as the product availabilities at the distribution centers and hospitals, and we move on to period $t+1$, resampling another $|\mathbb{S}|$ scenarios. In this way, we ensure that a sufficient number of scenarios is sampled for each time period and each realized set of initial conditions. Essentially, this is a simulation framework in which we re-optimize at each time period. The approach is very similar to a rolling horizon or model predictive control framework, except that as the time period advances, we keep the end of the time horizon fixed.

Table 3 summarizes the objective values of the two methods under the folding horizon framework. We applied the model to instances with 3 and 5 hospitals and one distribution center in 5 time periods. We assumed that the demand is still distributed uniformly in $\{0, \ldots, 5\}$. We set $c_{k1i} = 10i^2$, $b_{k11} = 2.5|\mathbb{I}|$, and $\zeta_{kit} = 3c_{k1i}$, for all $t \in \mathbb{T}$ as similar to the instances of Table 2. The results given for the proposed model each represent an average over 10 replications of the folding horizon framework. As seen in Table 3, the proposed method gives an improvement of roughly 2–5% versus the present approach. This is directly due to the fact that the model accounts for stochasticity of the demand and keeps the items at hospitals if needed. Moreover,

Table 3 Optimal objective value comparison of proposed and present methods under folding horizon framework, $\mathbb{T} = 5$

Scenarios	Number of hospitals					
	$\mathbb{I} = 3$			$\mathbb{I} = 5$		
	Proposed	Present	Reduction (%)	Proposed	Present	Reduction (%)
$\mathbb{S} = 500$	6125	6350	3.54	7060	7420	4.85
$\mathbb{S} = 1000$	6210	6350	2.19	7145	7420	3.71

the improvement versus the present approach seems to increase as the number of hospitals increases.

Unfortunately, the folding horizon approach is computationally expensive because we solve the problem iteratively and use only the result for the current time period. Therefore, the remaining experiments use the "single-shot" SAA approach, as in Table 2.

Demand correlation is of interest due to the clustering of surgeries in some time periods. In the next experiment, we assume the demands are autoregressive in order to model correlation over time and to explore the effect of this correlation on both the present and the proposed approaches. The motivation behind using autoregressive demands is that hospitals attempt to schedule similar surgeries in consecutive slots, e.g., on days when the surgeon is on duty, and also that hospitals sometimes postpone groups of surgeries simultaneously due to emergencies, operating room conflicts, or surgeon absences.

In particular, we assume an $AR(1)$ demand model, similar to [27, 33], follows:

$$d_{kist} = C + \gamma d_{kist-1}, \text{ for all } t = 2, ..., \mathbb{T}, \ 0 < \gamma < 1 \tag{10}$$

where d_{kis1} is a discrete random variable. Autoregressive models are known to be covariance stationary [26], i.e., the autocovariance does not vary over time, and the model requires only the first moment. Here, $\gamma = 1 - \frac{C}{E(d_{kis1})}$ ensures that the process mean does not change over time. The parameter γ is responsible for the dependency between demands.

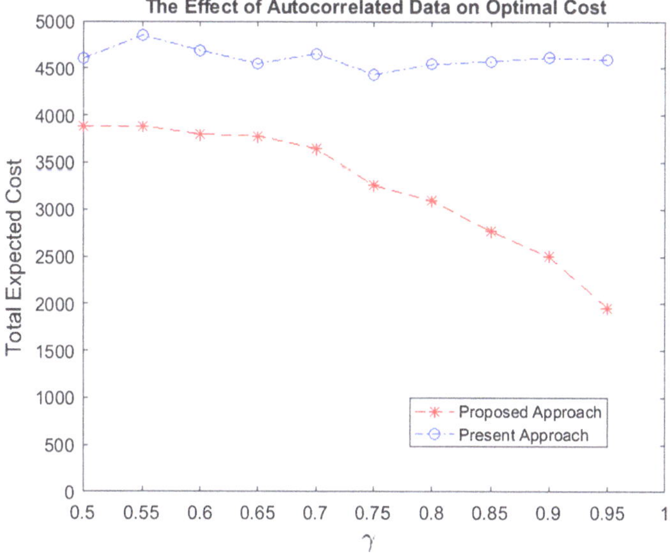

Fig. 2 The effect of autocorrelated data on optimal cost

Figure 2 illustrates the impact of γ in terms of total expected cost for the present and proposed approaches. We consider a one-week horizon ($\mathbb{T} = 7$) with one distribution center and three hospitals. Transportation and penalty cost coefficients are the same as their values in Table 2 for the case of $\mathbb{I} = 3$. We drew d_{kis1} from $U\{0, 5\}$ and set $C = E(d_{kis1})(1 - \gamma)$ in order for the model to retain its mean at each stage. We used the same stream of random numbers when testing the two approaches. Each point in Fig. 2 represents the total expected cost of one of the approaches for a give value of γ. Clearly, the present approach results in solutions that are more expensive than the proposed approach. Moreover, the present approach cannot take advantage of the autocorrelation, because it optimizes the distribution in each time period without considering future time periods. In contrast, our approach is able to leave kits at hospitals when they will be needed for future surgeries, thus reducing the cost. As γ increases, so does the clustering of surgeries over time, and so does the gap in cost between our approach and the present approach. It should be mentioned that by increasing the value of γ, the comparison in Fig. 2 is primarily based on realization of the demands for the first few time periods. In other words, for $\gamma \approx 1$, the demands are almost known in advance and therefore, the gap between the proposed and present approaches is largely due to the perfect knowledge of proposed model about the future demands.

To reduce the effect of γ and instead focus on the availability of kits, we investigate demand realizations in which the two models incur almost the same costs. In other words, we consider the effect of reducing the kit availability, v_k, on the two approaches, starting from the same cost. To do that, first, we set the demands d_{kist}

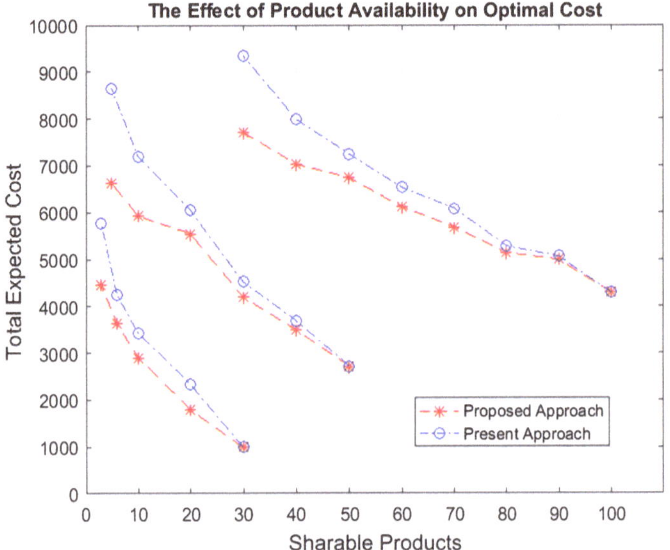

Fig. 3 The effect of product availability on optimal cost

sufficiently low, so that both approaches arrive at the same solution, namely, to bring the kits back to the DC after every use. Then, we gradually reduced v_k and solved the problems for each new value of v_k. From the figure, one can see that as the number of available kits decreases, the value of our proposed method versus the current approach increases. Figure 3 illustrates three instances, with one DC and 3, 5, and 10 hospitals. We used $c_{k1i} = 2i^2$, $\zeta_{kit} = 10c_{k1i}$, $t \in \mathbb{T}$, and assumed all scenarios are equally weighted. In addition, we used $v_k = 30$, $v_k = 50$, and $v_k = 100$ for all $k \in \mathbb{K}$ for the three instances, respectively. In this setting, the proposed method works better in terms of total cost. This advantage is due to the fact that the proposed model sees the demand and attempts to utilize the shareability property in the entire horizon rather than sending and receiving each item.

5 Optimization of Production Decisions

In this section, we extend our model to incorporate the decision of how many of the shareable products to manufacture; that is, we allow v_k to be a decision variable. We assume that the manufacturer produces the items at the beginning of the planning horizon and that there is no constraint on manufacturing capacity. Furthermore, we assume that inventory holding costs are negligible compared to production costs. Therefore, the total expected cost of production and distribution can be written as

$$T_{Cost} = \sum_k C_k v_k + \mathbb{Z}(v_k), \tag{11}$$

where C_k is the per-unit cost to manufacture a type-k implant kit and $\mathbb{Z}(v_k)$ is the optimal objective value of the multistage stochastic problem given above, assuming that the planning horizon begins with v_k available kits.

Figure 4 depicts the relationship among the production cost, the demand distribution, and the optimal cost for an instance with one distribution center, 3 hospitals, and 10 time periods. In particular, the demand follows a discrete uniform distribution on $\{0, \bar{d}\}$, where $\bar{d}$ is a parameter. We assume the transportation costs to the three hospitals are 100, 90, and 80, and that the unmet-demand penalty is 500.

In Fig. 4, there are two portions that are flat with respect to the z-axis; these are marked as (1) and (2). When the demand stays low relative to the production level, the model uses all kits at the start of the horizon and avoids any stockouts. Consequently, the flat areas represent the fact that the manufacturer has overproduced for the given amount of demand. The dashed red line on the figure indicates the optimal production level for each value of $\bar{d}$; that is, the minimizer of each level curve obtained by slicing along the $\bar{d}$-axis. In particular, if $\bar{d}$ is small, e.g., in (0, 3), the optimal production level remains roughly the same, up to a point at which the model realizes the necessity of producing new items. Then, for moderate values of $\bar{d}$, e.g., in (3, 6), the model becomes extremely sensitive toward increases in the maximum demand. Finally, for large $\bar{d}$, e.g., in (6, 10), the model again becomes almost indifferent toward

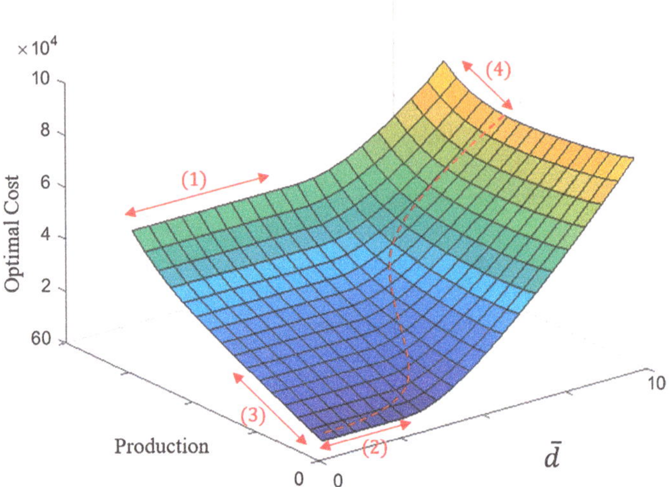

Fig. 4 Optimal cost of production–distribution model. Dashed red line (---) indicates line of optimal production level

increases in $\bar{d}$. Accordingly, there exist some demand intervals for which the model produces approximately the same number of items. Therefore, manufacturers in sharing system networks can use prior knowledge about the demand structure to avoid excess production, especially when production increases produce no improvement in the optimal cost (sections (3) and (4) in the figure).

6 Conclusion

In this chapter, we investigate distribution systems for shareable items using multistage stochastic optimization. We focus our study on artificial hip and knee joints, but the model could be adapted for other types of systems. We study a closed-loop supply network and formulate a model to minimize transportation and stockout costs. DC–hospital loops are used in the model to rebalance inventory levels and allow products to be reused. The random demands were modeled using discrete scenarios.

We solved several instances using AMPL and CPLEX and evaluated their elapsed CPU time and optimal objective function values. The results suggest that the proposed method outperforms the present approach used by our industry partner in terms of expected cost. The main reason for this is that the multistage model captures the possible future states, whereas the present approach is myopic and plans only for the current period. In addition, we evaluated the effect of problem input sizes on the elapsed CPU time; the results suggest that the model is most sensitive to increases in the number of scenarios.

Our model allows supply networks to take advantage of sharing systems to limit production levels. Sharing systems with closed production–distribution networks are promising types of supply chains. Tackling demand uncertainty in such supply chains using other approaches such as robust and/or chance-constrained optimization could be interesting areas for future research. Another could be the introduction of new elements into the model, such as lead times and usage times of the shareable parts.

References

1. Anvari, S., Turkay, M.: The facility location problem from the perspective of triple bottom line accounting of sustainability. Int. J. Prod. Res. **55**(21), 6266–6287 (2017)
2. Bachmeier, C.J.M., March, L.M., Cross, M.J., Lapsley, H.M., Tribe, K.L., Courtenay, B.G., Brooks, P.M., Arthritis Cost, Outcome Project Group.: A comparison of outcomes in osteoarthritis patients undergoing total hip and knee replacement surgery. Osteoarthr. Cartil. **9**(2), 137–146 (2001)
3. Ben-Tal, A., El Ghaoui, L., Nemirovski, A.: Robust Optimization. Princeton University Press, Princeton (2009)
4. Bertsimas, D., Thiele, A.: A robust optimization approach to supply chain management. In: International Conference on Integer Programming and Combinatorial Optimization, pp. 86–100. Springer, Berlin (2004)
5. Beswick, A.D., Wylde, V., Gooberman-Hill, R., Blom, A. and Dieppe, P.: What proportion of patients report long-term pain after total hip or knee replacement for osteoarthritis? A systematic review of prospective studies in unselected patients. BMJ Open **2**(1), e000435 (2012)
6. Birge, J.R., Louveaux, F.: Introduction to Stochastic Programming. Springer Science & Business Media, New York (2011)
7. Brinkmann, J., Ulmer, M.W., Mattfeld, D.C.: Inventory routing for bike sharing systems. Transp. Res. Procedia **19**, 316–327 (2016)
8. Carter, C.R., Ellram, L.M.: Reverse logistics: a review of the literature and framework for future investigation. J. Bus. Logist. **19**(1), 85 (1998)
9. Chemla, D., Meunier, F., Calvo, R.W.: Bike sharing systems: solving the static rebalancing problem. Discret. Optim. **10**(2), 120–146 (2013)
10. Chen, Z., Yan, Z.: Scenario tree reduction methods through clustering nodes. Comput. Chem. Eng. **109**, 96–111 (2018)
11. Chiariotti, F., Pielli, C., Zanella, A., Zorzi, M.: A dynamic approach to rebalancing bike-sharing systems. Sensors **18**(2), 512 (2018)
12. Chow, J.Y.J., Sayarshad, H.R.: Symbiotic network design strategies in the presence of coexisting transportation networks. Transp. Res. Part B Methodol. **62**, 13–34 (2014)
13. Correia, I., Nickel, S., Saldanha-da Gama, F.: A stochastic multi-period capacitated multiple allocation hub location problem: formulation and inequalities. Omega **74**, 122–134 (2018)
14. Cruz, F., Subramanian, A., Bruck, B.P., Iori, M.: A heuristic algorithm for a single vehicle static bike sharing rebalancing problem. Comput. Oper. Res. **79**, 19–33 (2017)
15. De Brito, M.P., Dekker, R.: A framework for reverse logistics. In: Reverse Logistics, pp. 3–27. Springer, Berlin (2004)
16. Defourny, B., Ernst, D., Wehenkel, L.: Multistage stochastic programming: a scenario tree based approach. In: Decision Theory Models for Applications in Artificial Intelligence: Concepts and Solutions, p. 97 (2011)
17. Devika, K., Jafarian, A., Nourbakhsh, V.: Designing a sustainable closed-loop supply chain network based on triple bottom line approach: a comparison of metaheuristics hybridization techniques. Eur. J. Oper. Res. **235**(3), 594–615 (2014)

18. Ding, T., Yuan, H., Bie, Z.: Multi-stage stochastic programming with nonanticipativity constraints for expansion of combined power and natural gas systems. IEEE Trans. Power Syst. **33**(1), 317–328 (2018)
19. Ethgen, O., Bruyere, O., Richy, F., Dardennes, C., Reginster, J.-Y.: Health-related quality of life in total hip and total knee arthroplasty: a qualitative and systematic review of the literature. JBJS **86**(5), 963–974 (2004)
20. Gharehyakheh, A., Tavakkoli-Moghaddam, R.: A fuzzy solution approach for a multi-objective integrated production-distribution model with multi products and multi periods under uncertainty. Manag. Sci. Lett. **2**(7), 2425–2434 (2012)
21. Gharehyakheh, A., Cantu, J., Rogers, K.J.: A systematic review of quantitative modeling approach in sustainable supply chain under uncertainty. In: Proceedings of the International Annual Conference of the American Society for Engineering Management, pp. 1–7. American Society for Engineering Management (ASEM), Huntsville (2017)
22. Golroudbary, S.R., Zahraee, S.M.: System dynamics model for optimizing the recycling and collection of waste material in a closed-loop supply chain. Simul. Model. Pract. Theory **53**, 88–102 (2015)
23. Haddadsisakht, A., Ryan, S.M.: Closed-loop supply chain network design with multiple transportation modes under stochastic demand and uncertain carbon tax. Int. J. Prod. Econ. **195**, 118–131 (2018)
24. Hooshmand, F., MirHassani, S.A.: Reduction of nonanticipativity constraints in multistage stochastic programming problems with endogenous and exogenous uncertainty. Math. Methods Oper. Res. **87**(1), 1–18 (2018)
25. Kazemian, I., Aref, S.: Multi-echelon supply chain flexibility enhancement through detecting bottlenecks. Glob. J. Flex. Syst. Manag. **17**(4), 357–372 (2016)
26. Kilian, L., Lütkepohl, H.: Structural Vector Autoregressive Analysis. Cambridge University Press, Cambridge (2017)
27. Kovtun, V., Giloni, A., Hurvich, C.: Aggregated information in supply chains (2018)
28. Laumanns, M., Woerner, S.: Multi-echelon supply chain optimization: methods and application examples. In: Optimization and Decision Support Systems for Supply Chains, pp. 131–138. Springer, Cham (2017)
29. Li, S., Luo, Q., Hampshire, R.: Design of multimodal network for mobility-as-a-service: first/last mile free floating bikes and on-demand transit (2017)
30. Özceylan, E., Paksoy, T.: A mixed integer programming model for a closed-loop supply-chain network. Int. J. Prod. Res. **51**(3), 718–734 (2013)
31. Pagnano, M., Cushner, F.D., Hansen, A., Scuderi, G.R., Scott, W.N.: Blood management in two-stage revision knee arthroplasty for deep prosthetic infection. Clin. Orthop. Relat. Res. **367**, 238–242 (1999)
32. Parlar, M., Perry, D.: Inventory models of future supply uncertainty with single and multiple suppliers. Nav. Res. Logist. (NRL) **43**(2), 191–210 (1996)
33. Pirhooshyaran, M., Niaki, S.T.A.: A double-max MEWMA scheme for simultaneous monitoring and fault isolation of multivariate multistage auto-correlated processes based on novel reduced-dimension statistics. J. Process. Control. **29**, 11–22 (2015)
34. Pishvaee, M.S., Jolai, F., Razmi, J.: A stochastic optimization model for integrated forward/reverse logistics network design. J. Manuf. Syst. **28**(4), 107–114 (2009)
35. Raviv, T., Kolka, O.: Optimal inventory management of a bike-sharing station. IIE Trans. **45**(10), 1077–1093 (2013)
36. Rezaee, A., Dehghanian, F., Fahimnia, B., Beamon, B.: Green supply chain network design with stochastic demand and carbon price. Ann. Oper. Res. **250**(2), 463–485 (2017)
37. Schuijbroek, J., Hampshire, R.C., Van Hoeve, W.-J.: Inventory rebalancing and vehicle routing in bike sharing systems. Eur. J. Oper. Res. **257**(3), 992–1004 (2017)
38. Shankar, R., Bhattacharyya, S., Choudhary, A.: A decision model for a strategic closed-loop supply chain to reclaim end-of-life vehicles. Int. J. Prod. Econ. **195**, 273–286 (2018)
39. Shapiro, A., Philpott, A.: A tutorial on stochastic programming. Manuscript. www2.isye.gatech.edu/ashapiro/publications.html (2007)

40. Sinha, R.K.: Hip Replacement: Current Trends and Controversies. CRC Press, Boca Raton (2002)
41. Snyder, L.V.: Facility location under uncertainty: a review. IIE Trans. **38**(7), 547–564 (2006)
42. Snyder, L.V., Daskin, M.S.: Stochastic p-robust location problems. IIE Trans. **38**(11), 971–985 (2006)
43. Snyder, L.V., Shen, Z.-J.M.: Fundamentals of Supply Chain Theory. Wiley, New York (2011)
44. Soleimani, H., Govindan, K.: Reverse logistics network design and planning utilizing conditional value at risk. Eur. J. Oper. Res. **237**(2), 487–497 (2014)
45. Spahn, D.R.: Anemia and patient blood management in hip and knee surgerya systematic review of the literature. Anesth. J. Am. Soc. Anesth. **113**(2), 482–495 (2010)
46. Spiliotopoulou, E., Donohue, K., Gürbüz, M.Ç., Heese, H.S.: Managing and reallocating inventory across two markets with local information. Eur. J. Oper. Res. **266**(2), 531–542 (2018)
47. Srivastava, S.K.: Network design for reverse logistics. Omega **36**(4), 535–548 (2008)
48. Wiesemann, W., Kuhn, D., Sim, M.: Distributionally robust convex optimization. Oper. Res. **62**(6), 1358–1376 (2014)
49. Yi, P., Huang, M., Guo, L., Shi, T.: A retailer oriented closed-loop supply chain network design for end of life construction machinery remanufacturing. J. Clean. Prod. **124**, 191–203 (2016)

TopSpin: TOPic Discovery via Sparse Principal Component INterference

Martin Takáč, Selin Damla Ahipaşaoğlu, Ngai-Man Cheung and Peter Richtárik

Abstract We propose a novel topic discovery algorithm for unlabeled images based on the bag-of-words (BoW) framework. We first extract a dictionary of visual words and subsequently for each image compute a visual word occurrence histogram. We view these histograms as rows of a large matrix from which we extract sparse principal components (PCs). Each PC identifies a sparse combination of visual words which co-occur frequently in some images but seldom appear in others. Each sparse PC corresponds to a topic, and images whose interference with the PC is high belong to that topic, revealing the common parts possessed by the images. We propose to solve the associated sparse PCA problems using an Alternating Maximization (AM) method, which we modify for the purpose of efficiently extracting multiple PCs in a deflation scheme. Our approach attacks the maximization problem in SPCA directly and is scalable to high-dimensional data. Experiments on automatic topic discovery

This work was partially supported by the U.S. National Science Foundation, under award number NSF:CCF:1618717, NSF:CMMI:1663256 and NSF:CCF:1740796.

M. Takáč (✉)
Lehigh University, Bethlehem, PA 18015, USA
e-mail: Takac.MT@gmail.com

S. D. Ahipaşaoğlu · N.-M. Cheung
Singapore University of Technology and Design, Singapore 487372, Singapore
e-mail: ahipasaoglu@sutd.edu.sg

N.-M. Cheung
e-mail: ngaiman_cheung@sutd.edu.sg

P. Richtárik
King Abdullah University of Science and Technology (KAUST), Thuwal, Kingdom of Saudi Arabia
e-mail: peter.richtarik@kaust.edu.sa; peter.richtarik@ed.ac.uk

P. Richtárik
University of Edinburgh, Edinburgh, UK

P. Richtárik
Moscow Institute of Physics and Technology, Dolgoprudny, Russia

J. D. Pintér and T. Terlaky (eds.), *Modeling and Optimization: Theory and Applications*, Springer Proceedings in Mathematics & Statistics 279, https://doi.org/10.1007/978-3-030-12119-8_8

and category prediction demonstrate encouraging performance of our approach. Our SPCA solver is publicly available.

Keywords Sparse PCA · Bag-of-words · Topic discovery · Hidden topic

1 Introduction

The goal of this paper is to design a method performing the following: *Given a database of n images, identify k (not necessarily disjoint) collections of images,* $S_1, \ldots, S_k$, *each corresponding to a certain "topic"*. Definition of a topic is not provided, and hence we are looking for an *unsupervised* learning method able to first (i) automatically identify the *hidden* topics from the images, then to (ii) form collections of images belonging to these topics, and finally, (iii) if a new image is presented, find the topics related to this image and place this new image into the suitable collection(s). Unsupervised topic discovery has been a popular topic in the recent years (e.g., see [1, 8, 10, 17]). It is a challenging topic that usually requires machinery such as directed graphical models, time series modeling, Hierarchical Bayesian methods, nonparametric Bayesian methods, model selection, and mixed membership models. Here, we present a simple approach to tackle topic discovery that relies on linear algebra and optimization, bypassing any discussion on the underlying probabilistic model.

For instance, consider a database of photos with people, cars, and buildings on them (without knowing this). Some photos may contain people and no cars nor buildings, some may have people and cars, some may be photos of buildings unspoiled by cars or people. From the viewpoint of the "cars" topic, people and buildings are clutter/background. From the viewpoint of the "people" topic, cars and buildings are background and not essential. We would wish to be able to automatically discover these three topics and identify three collections. If there are images with both cars and people, then these images will be assigned to two topics instead of only one. On the other hand, it may be that people and cars always occur together in an image, while people and buildings also always occur together. In that case, the topics which we would wish to discover are "people and cars" and "people and buildings".

It has recently been demonstrated [21] that sparse PCA is able to discover topics in a database of articles. The approach is applied to a data-matrix A where rows correspond to articles, columns to words and $A_{i,j}$ is equal to the frequency of word j in article i. For example, [21] showed that in a NYTimes article dataset, words associated with the first and second sparse PCs have nonzero weights on words *million, percent, business, company, market, companies* and *point, play, team, season, and game*, respectively. These words discover two of the most important topics in the articles: business and sports.

One of our contributions is to show that a similar approach can be successfully applied to images. As we shall see, identification of topics in image databases can be performed by extracting *sparse principal components* of a matrix whose rows correspond to *all* images in the database, columns to visual words (obtained by quantization of local descriptors such as SIFT, via clustering), with the (i, j) entry representing the frequency of visual word j in image i. Images are subsequently assigned to the identified topics using a simple technique we call *interference*: images whose interference with a PC is high form natural topics.

Contents: We start in Sect. 2 by briefly reviewing some of the relevant literature. In Sect. 3, we propose and describe TopSpin, an algorithm for topic discovery. Further, in Sect. 4 we provide some background for sparse PCA and present a scalable algorithm for extracting sparse PCs. In Sect. 5, we provide numerical evidence for the efficacy and efficiency of our approach. Finally, we conclude in Sect. 6 with a brief summary of our main contributions.

2 Literature Review

In the unsupervised visual object categorization problem, we attempt to uncover the category information of an image dataset without relying on any information capturing image content [1, 8, 10, 17]. Unsupervised categorization relieves the burden of human labeling and removes subjective bias. Grauman and Darrell [8] proposed a graph-based method for unsupervised object categorization. In their work, the sets of local feature descriptors extracted from individual database images are graph nodes, while graph edges are weighted by the number of correspondences between images. A spectral clustering algorithm is then applied to the graph's affinity matrix to produce image groupings. Sivic et al. [17] demonstrated unsupervised learning of object hierarchy from datasets of unlabeled images. In their work, the generative Hierarchical Latent Dirichlet Allocation (hLDA) model, previously used for text analysis [2], is adapted to the visual domain. Images are represented by a visual vocabulary of quantized SIFT descriptors. A "coarse-to-fine" description of the images with varying degrees of appearance and spatial localization granularity is proposed to facilitate discovery of visual object class hierarchies. Bart et al. [1] also proposed unsupervised learning of visual taxonomies independent of Sivic et al. [17]. They use a modified nonparametric prior over tree structure of a certain depth [2]. Their modified model allows to represent several topics at each node in the taxonomy and makes available all topics at every node to facilitate visual taxonomies inference. Images are represented using space-color histograms. Based on the BoW framework, Kinnunen et al. [10] applied the self-organization principle and the Kohonen map to solve unsupervised visual object categorization.

Our work is also related to object recognition. One important difference is that we do not assume any prior category information: as will be discussed, we discover object categories automatically from the dataset, and the testing images are assigned to these object categories using the same framework.

In object recognition, the use of local descriptors with high degree of invariance has become one of the dominant approaches [20]. In particular, in the BoW approach, an image is represented by a bag of highly invariant local feature descriptors (e.g., [11]). These local descriptors may be further clustered or quantized into a dictionary of visual words [18]. A visual word occurrence histogram of an image is used to determine a distance function for classification of object categories. To generate a large dictionary of vocabularies, hierarchical quantization can be used to produce a vocabulary tree with the leaf nodes being the visual words [15]. A recent work of Naikal et al. [14] used Sparse PCA to select informative visual words to improve object recognition. Given the prior object category information, they apply Sparse PCA to each object category separately to select informative (more useful) visual words within individual categories. The union of all the informative visual words selected from individual categories forms the overall refined visual dictionary. Different from Naikal et al. [14], our work discovers object categorization automatically by applying Sparse PCA in a different way (and with different philosophy). We propose to perform category prediction by projecting the test image's occurrence histogram vector directly onto the principal components (PCs) associated with the discovered categories, and this is different from previously proposed BoW-based object recognition systems. We argue that with our approach each PC selects and associates co-occurring visual words that are signatures for a category. The projection of the test image's histogram onto a PC quantifies the extent of visual words co-occurrence in the test image, which is useful for predicting the category.

3 Topic Discovery Algorithm

We propose TopSpin (Algorithm 1), a method for TOPic discovery via Sparse Principal component INterference.

Algorithm 1 TopSpin

Input: n images, p=#visual words, k=#topics, s=sparsity

1. **Representation:**
   1a. Represent each image i by a row vector $h^i \in \mathbb{R}^p$
   1b. Compute weight vector $w \in \mathbb{R}^p$, $w \geq 0$
2. **Extract topics via sparse PCA:**
   2a. Form $A = H\,\mathrm{Diag}(w) \in \mathbb{R}^{n \times p}$, where $H_{i:} = h^i$
   2b. Extract s-sparse PCs $x^1, \ldots, x^k \in \mathbb{R}^p$ from A
3. **Detect topic images via Interference:**
   3a. Choose topic threshold values $\delta_1, \ldots, \delta_k > 0$
   3b. $S_l \leftarrow \{i \;:\; \mathrm{Intf}(i, x^l) > \delta_l\}$, $l = 1, \ldots, k$

Output: S_l (images associated with topic l), $l = 1, \ldots, k$

3.1 Step 1: Representation

In *Step 1a*, we utilize the standard Bag-of-Words (BoW) approach, where for each image, we identify keypoints (e.g., by Maximally Stable Extremal Regions (MSER)), and then find local feature descriptors for them (e.g., by SIFT algorithm; SIFT descriptors are 128-dimensional vectors). We identify a high number of descriptors for each image and then select a random subset and perform clustering, obtaining p cluster centers ("visual words"). Each local descriptor in an image is then substituted by the closest visual word (distances are measured in L_2 norm). Therefore, image i can be described by a histogram vector $f^i \in \mathbb{R}^p$ as follows: f^i_j is the number of appearances of visual word j in image i. For normalization purposes (e.g., sharpness, size) we instead represent each image i by the *normalized histogram* $h^i = f^i/\sum_j f^i_j$. While in this paper we focus on this particular image representation, our framework also applies to other representations.

Some visual words may be more important than others. For instance, a word appearing in all images with identical frequency is not informative and hence can be excluded from further analysis. In *Step 1b*, we associate with each visual word $j = 1, 2, \dots, p$ a weight $w_j \geq 0$, forming a vector $w \in \mathbb{R}^p_+$. In the experiments in this paper, we work with the Term Frequency Inverse Document Frequency (tf-idf) weights [15] defined by $w_j = \ln(n/n_j)$, where $n_j = |\{i : h^i_j > 0\}|$, i.e., the number of images containing visual word j. If word j occurs in many images, then w_j is small and vice versa. However, different weights might be preferable depending on the dataset.

3.2 Step 2: Topic Extraction

In this step, we extract k leading sparse principal components (sparse PCs) of the matrix $A = H\,\mathrm{Diag}(w)$, where the i-th row of $H \in \mathbb{R}^{n\times p}$ is h^i and $\mathrm{Diag}(w)$ is the $p \times p$ diagonal matrix with vector w on the diagonal. Various sparse PCA formulations were suggested in the literature. Here, we propose the s-sparse PC x^l to be obtained as the solution of the following optimization problem:

$$\text{maximize } \|A_l x\|_2^2 \text{ subject to } \|x\|_2 \leq 1,\ \|x\|_0 \leq s, \tag{1}$$

where $\|\cdot\|_2$ is the standard Euclidean norm, $\|x\|_0 = |\{i : x_i \neq 0\}|$ (number of nonzero elements in x), and $A_{l+1} = A_l - x^l(x^l)^T$ with $A_1 = A$. Further, we propose that (1) be solved by the simple yet powerful Alternating Maximization (AM) framework [16]. As will be discussed, the proposed method is scalable, fast and can be run in parallel on multicore machines, GPUs and clusters.[1] We further propose an efficient solution of a sequence of problems (1) for $l = 1, 2, \dots, k$ (deflation techniques for sparse PCA are described in [12]). A naive approach would be to simply

[1] Source code: https://code.google.com/p/24am/.

solve (1) in a loop, forming A_{l+1} from A_l as described above. However, this is not efficient due to the structure and sparsity of the problem. We therefore implement our own multicore version of the method in C++ suitable for the task. Our SPCA solver is three orders of magnitude faster than the Augmented Lagrangian Method (ALM) proposed by [14] for $p = 500$ and its advantage grows with p. More details on SPCA, AM, an extension of AM, and a comparative study between AM and ALM are given in Sect. 4.

3.3 Step 3: Topic Detection

Define the *interference* between PC x^l and image i via

$$\mathrm{Intf}(i, x^l) := \Phi(\sum_{j=1}^{p} h^i_j w_j x^l_j), \tag{2}$$

where Φ is some function. When $\Phi(\cdot) = |\cdot|$, it corresponds to the absolute value of the inner/dot product between x^l and $a^i := (h^i_1 w_1, h^i_2 w_2, \ldots, h^i_p w_p)^T$ (the i-th row of A). In this case, $\mathrm{Intf}(i, x^l)$ is in fact the length of the projection of a^i onto x^l: it quantifies the extent that image i contains the visual words associated with PC x^l. In *Step 3b*, we define S_l to be the set of images i having large enough interference with x^l, where the precise quantitative meaning of "large enough" is controlled by the parameter δ_l chosen in *Step 3a*. As we shall see from computational experiments (for instance, see Fig. 5), images having high interference with a PC indeed belong to the same topic/category.

3.4 An Illustrative Example

We illustrate the method on a simplified artificial example (see Fig. 1). We have $n = 9$ images which naturally belong to 3 categories/topics: guns, mice, and bicycles. In *Step 1*, we identify 8 visual words: 3 for guns (green, brown and pink dots), 2 for mice (blue and dark green dots) and 3 for bicycles (light blue, purple and orange dots). In this case, the situation is perfect as no two images in different topics contain the same visual word. Here, we choose w to be the vector of all ones. As a consequence, A is block diagonal, with rows $a^1, \ldots, a^9$ as depicted in *Step 2* in Fig. 1. In *Step 2* of TopSpin, sparse PCs x^1, x^2 and x^3 are computed (we set $s = 3$). Each sparse PC has zero values at the coordinates that correspond to the visual words that are related to two topics and nonzero values for visual words that are related to a single topic. In this sense, each sparse PC (perfectly) identifies a topic. In particular, x^1 represents the "mice" topic, x^2 represents the"bicycles" topic and x^3 represents the "guns" topic. Finally, in *Step 3* for each x^l, we compute the interferences with each

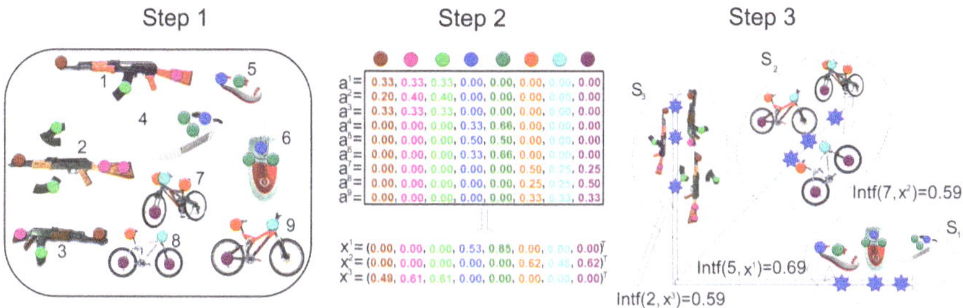

Fig. 1 Illustration of the three steps of the TopSpin method

normalized histogram vector a^i. The last step in Fig. 1 plots each image in a 3D space, with the coordinates of image i being $(\text{Intf}(i, x^1), \text{Intf}(i, x^2), \text{Intf}(i, x^3))$. In this example, the interferences of i with x^l will be nonzero if and only if i belongs to the topic represented by PC x^l. Hence, each of the sets S_l, $l = 1, 2, 3$, will consist of images depicted on a single axis in the 3D space. The three sets S_1, S_2, S_3 identified by TopSpin correspond perfectly to the natural topics inherent in the image database.

Real data sets are different from the simplified example depicted in Fig. 1 in several ways. First, there will be many images and many visual words. Second, A will not be block diagonal—images will naturally share visual words with other images since they may share multiple objects. As a consequence, the topics discovered by TopSpin will not be perfect as in the simplified example. Please see Sect. 5 for numerical experiments with real datasets.

4 Sparse Principal Component Analysis

Principal Component Analysis (PCA) is an important tool for dimension reduction and data analysis. Let $A \in \mathbb{R}^{n \times p}$ denote a data matrix where the rows correspond to measurements of p variables. PCA finds linear combinations of the columns of A, called principal components (PCs), pointing in mutually orthogonal directions, together explaining as much variance in the data as possible. If the rows of A are centered, the problem of extracting the first PC can be written as $\max\{\|Ax\| : \|x\|_2 \leq 1\}$, where $\|\cdot\|$ is any norm for measuring variance.[2] Although classical PCA employs the L_2 norm, L_1 norm can also be used—this is especially useful when the data is contaminated (e.g., by outliers). Further, PCs can be obtained by deflation as explained in the previous section.

PCA usually produces PCs that are combinations of *all* variables. In many applications however, including topic discovery, it is desirable to induce sparsity into the PCs. The problem of finding PCs with few nonzero components is known as sparse PCA or SPCA (see [5, 6, 9, 22]). Sparsity is usually incorporated either directly by

[2]A simple scaling argument shows that the solution must satisfy $\|x\|_2 = 1$.

enforcing a constraint on the number of nonzero components in a PC, such as in (1), or by adding a penalty term to the objective function.

4.1 SPCA via Alternating Maximization

In general, calculation sparse PCs for large dataset is a challenging task. Various methods have been devised in the literature in recent years. These methods differ in the problem formulation (constrained versus penalized) and the approaches they follow in handling the nonconvex objective function (penalized formulations) or the nonconvex feasible region (constrained formulations). Usually, there is a convexification step in the process. The solutions obtained are not global maximizers, therefore several local optimal solutions are obtained by starting the algorithms at different initial solutions or introducing some randomization within the process. For example, [6, 14] work with an SDP relaxation of the penalized formulation and employ techniques from convex optimization to find solutions to the relaxed problem.

In our work, we propose to use an Alternating Maximization (AM) framework (described below in detail) to calculate the sparse PCs. Our framework is scalable and amendable for parallel implementations for various architectures. Thus, it is suitable for large scale PCA and SPCA problems, as will be demonstrated by experiment results. The basic principle of an AM algorithm is: for a certain function $F(x, y)$ to be maximized over compact sets X and Y, the two steps of an AM algorithm are of the following alternating maximization form [16]: $y = \arg\max_y\{F(x, y) : y \in Y\}$ and $x = \arg\max_x\{F(x, y) : x \in X\}$. In what follows, we discuss how we reformulate the sparse PCA problem into an appropriate form that can be attacked by AM.

In our reformulation, we start with the constrained formulation

$$\text{maximize } \|Ax\|_2^2 \text{ subject to } \|x\|_2 \leq 1, \ \|x\|_0 \leq s. \tag{3}$$

This can be reformulated as

$$\begin{aligned} \text{maximize} \quad & y^T Ax \\ \text{subject to} \quad & \\ & \|x\|_2 \leq 1, \ \|x\|_0 \leq s, \\ & \|y\|_2 \leq 1, \end{aligned}$$

by introducing dummy variables $y \in \mathbb{R}^n$ and making use of the following fact:

Observation 1 *The maximum of $y^T Ax$ over*

$$\{y \in \mathbb{R}^n \ : \ \|y\|_2 \leq 1\}$$

is equal to $\|Ax\|_2$ and the maximizer is attained at $y^* = \frac{Ax}{\|Ax\|_2}$.

Using this together with another fact (given below in Observation 2), we obtain the AM-SPCA method presented as Algorithm 2.

Observation 2 *The maximum of* $y^T Ax$ *over*

$$\{x \in \mathbb{R}^n \ : \ \|x\|_2 \leq 1 \ \|x\|_0 \leq s\}$$

is equal to $T_s(A^T y)$. *Moreover, the maximizer is attained at* $x^* = \frac{T_s(A^T y)}{\|T_s(A^T y)\|_2}$, *where* $T_s(a)$ *is a vector obtained from a by keeping the s largest elements in absolute value and setting the rest to zero. This is also referred to as the hard-thresholding operator.*

Algorithm 2 AM-SPCA: Alternating Maximization

Select initial point $x^{(0)} \in \mathbb{R}^p$ and $t \leftarrow 0$
Repeat
  $y^{(t)} = Ax^{(t)}/\|Ax^{(t)}\|_2$
  $x^{(t+1)} \leftarrow T_s(A^T y^{(t)})/\|T_s(A^T y^{(t)})\|_2$
  $t \leftarrow t+1$
Until a stopping criterion is satisfied

We implement a parallel version of the AM-SPCA, where we start the algorithm at many points simultaneously. This turns the matrix-vector multiplications in the main loop of the algorithm into matrix-matrix multiplications which are known to be advantageous in parallel architectures. This enables us to obtain very good quality solutions in short amount of time, much faster than the other algorithms devised in the literature.

To demonstrate the scalability of our devised AM-SPCA, in what follows, we compare our AM-SPCA framework with another state-of-the-art method that is recently developed within the image processing community.

4.2 *AM-SPCA Versus ALM*

ALM is an Augmented Lagrangian Method proposed in [14] for object recognition. It applies an SDP *relaxation* of the Sparse PCA formulation in [6]. ALM does not control the sparsity level of the solution directly, but via a penalty parameter the value of which is a very poor predictor of sparsity. If a particular target sparsity is sought, one needs to run ALM repeatedly with different values of the penalty parameter, effectively fine-tuning for it. On the other hand, our AM-SPCA does not suffer from this issue as sparsity is controlled directly by s. (Recall that we use the cardinality constrained formulation.)

In Fig. 2, we compare the performance of AM-SPCA and ALM on artificial random matrices $A \in \mathbb{R}^{n \times p}$ with $n = \frac{p}{2}$ and $p \in \{10, 50, \ldots, 500, \ldots, 10{,}000\}$. For

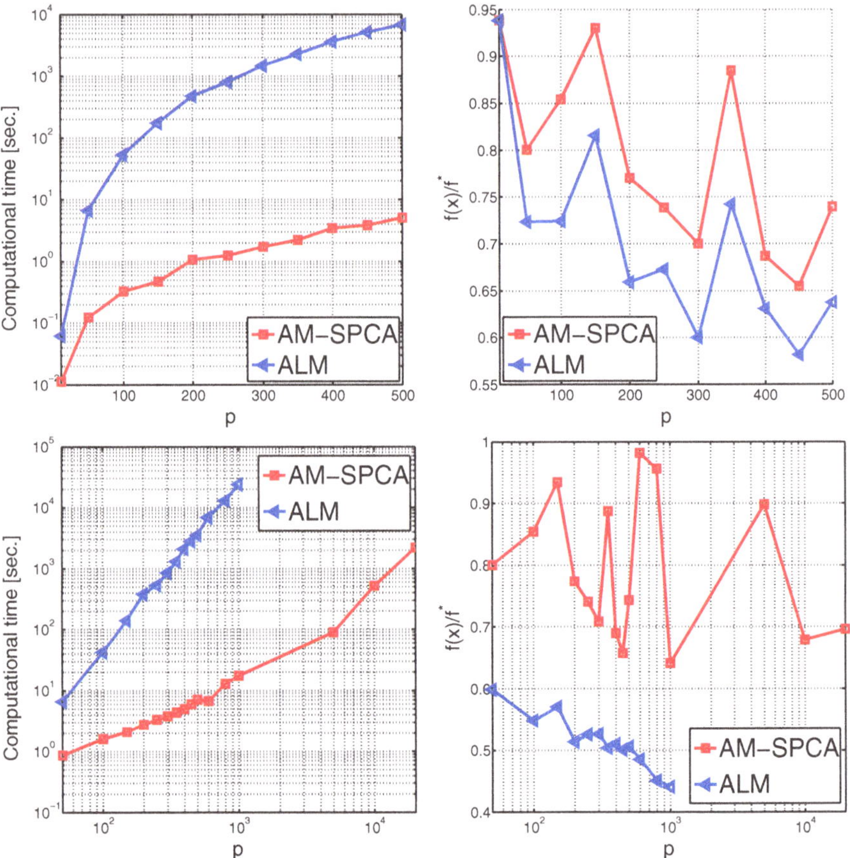

Fig. 2 The computational time and solution quality of AM-SPCA and ALM algorithms for small (upper) and large (lower) problem sizes. We have used a Linux machine with 24 cores (Intel Xeon X5650, 2.67GHz) and 24GB RAM. We have implemented both codes in C++ using Intel MKL and GSL BLAS library

each problem, we fixed a penalty parameter and obtained a single leading sparse PC using the ALM method. We then measured the resulting sparsity s of the solution. Subsequently, we run AM-SPCA with target sparsity level set to s. Here are our findings. First, AM-SPCA terminates *three orders of magnitude faster* than ALM for $p = 500$; with the gap getting *larger* with p (left plots). Hence, AM-SPCA is well suited for problems where it is beneficial to work with a large number of visual words. Second, AM-SPCA solutions for all problem instances are of *better quality* than those obtained by ALM (right plots) in the sense that they explain more of the optimal variance. That is, the ratio $f(x)/f^*$ is larger, where $f^* = \|Ax^*\|_2^2$ and x^* is the optimal non-sparse PC, and $f(x) = \|Ax\|_2^2$ and x is the s-sparse PC found by the methods.

5 Numerical Experiments

In this section we highlight, on a sequence of carefully chosen experiments, the efficacy and efficiency of TopSpin. We first work with the BMW (Berkeley Multi-view Wireless) dataset [13, 14] consisting of 20 image categories (Berkeley campus buildings), with $16 \times 5 = 80$ images in each. In each category, the same building is captured repeatedly from different distances and angles 16 times, each time simultaneously by 5 cameras attached to a fixed frame in close proximity to one another. Hence, there is a total of $1600 = 20 \times 16 \times 5$ images.

In this experiment, we use MSER keypoints and SIFT descriptors, and our codes were implemented in C++. We used OpenCV library v2.4.4.0 to find the keypoints and extract local descriptors and for hierarchical clustering (using FLANN) to obtain a dictionary of visual words. In the whole section, we have used $\Phi(\cdot) = \|\cdot\|$ for the computation of interference.

5.1 *Topic Discovery*

In this section, we empirically show that sparse PCs can identify topics. We took all images from camera #1 (320 images), used p =5,000 visual words and extracted PCs with $s = 20$.

For illustration purposes, we limit our attention to just 3 topics; the message applies to more topics of course. The top row of Fig. 3 depicts the interference between sparse PCs x^2, x^6 and x^7 and images belonging to three different topics/categories (red, blue and green). One can observe that indeed each sparse PCs have high interference with a single image topic. The next three rows show the same image four times; in the first column with all visual words, in the second column with only those visual words selected by x^2 (i.e., in the set $\{j : x_j^2 \neq 0\}$), in the third column only those visual words selected by x^6 and in the fourth column only those visual words selected by x^7. Clearly, x^2 selects a substantial number of visual words in the second row image and does not select nearly any visual words in the third and fourth row image. The top image has high interference with x^2, while bottom two images has low interference. The situation with x^6 and x^7 is reversed. Indeed, the top image belongs to S_2, the topic attached to x^2, the middle image belongs to S_6 while the bottom image belongs to S_7.

In Fig. 4, we focus on the same three categories as in the previous test, but in this case, we visualize them in 3D space, as in Fig. 1. Because each image is represented by a p = 5,000 dimensional vector (h^i), a naïve approach for visualizing h^i in 3D would be to project the vectors h^i onto a random 3D subspace of $\mathbb{R}^p$ (Fig. 4, left). No apparent separation of the images belonging to the three topics (represented by different color and marker) is present. However, if we project onto the space spanned by the PCs corresponding to the three topics, we can clearly see the images belonging to different topics coalescing around different axes (Fig. 4, right).

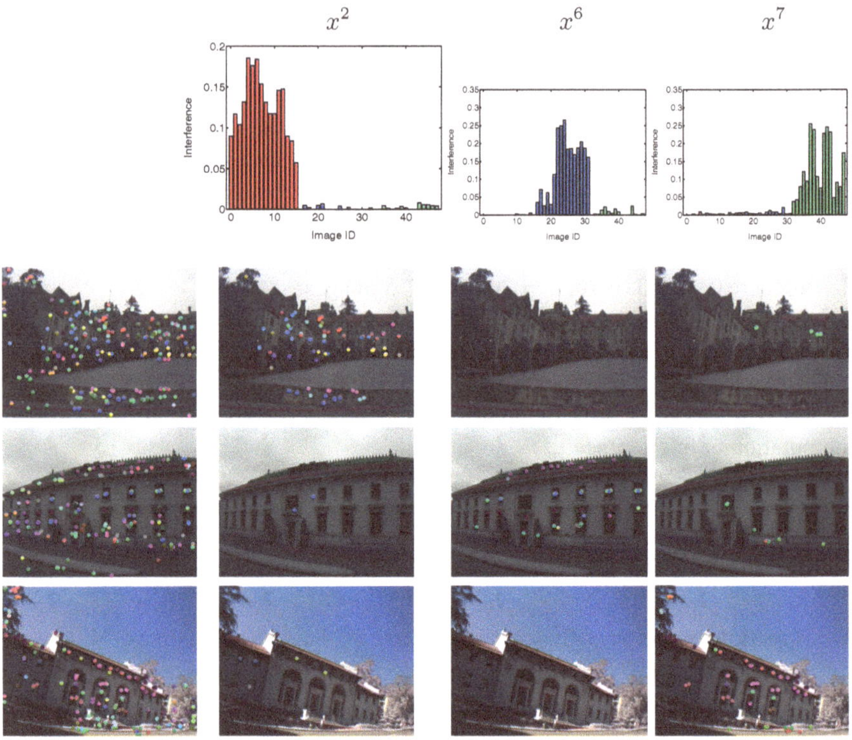

Fig. 3 Different principal components select visual words prevalent in different categories

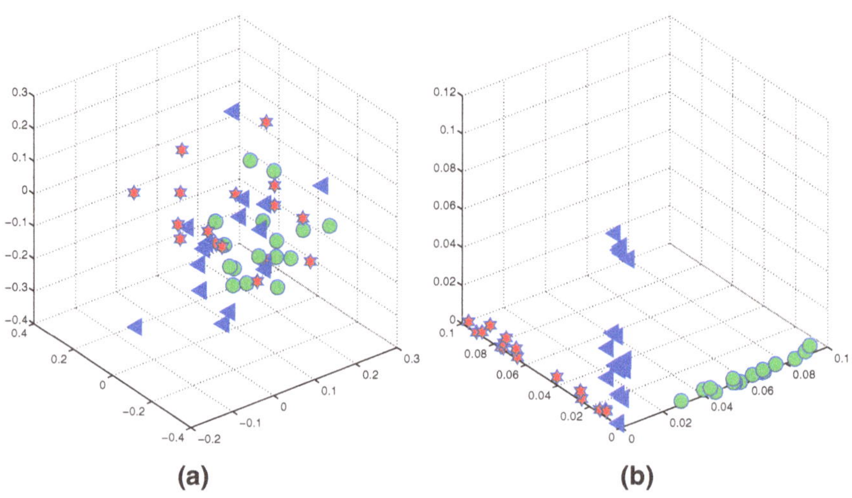

Fig. 4 Random vectors do not identify topics, sparse PCs do

Fig. 5 Each row represents one sparse PC: x^1 (row 1) and x^7 (row 2). The plots in the first column show the interferences of all images (x-axis) with the given PC. Images from the same category/topic (not known to our method!) are represented by the same color (but each color is used three times and each time it represents a different category/topic). For each PC, we show the 8 images having the largest interference with it; these are the sets S_1 and S_7 for appropriate choice δ_1 and δ_7. One can observe that not only TOP-SPIN selects important features (visual words), but the method correctly identifies topics

Let us now look at (a portion of) the actual output of TopSpin for $k = 7$, with a dictionary of size $p = 5{,}000$ and $s = 50$. Figure 5 depicts the sets S_1 and S_7 for δ_1 and δ_7 chosen so that $|S_1| = |S_7| = 8$. It is clear that the method is able to identify the categories. We would like to stress that Sparse PCA is applied to the *entire* training dataset, and that testing is done on different images. In contrast, the approach in [14] presupposes the knowledge of the categories as Sparse PCA is applied to test images from each category. As we shall see later, TopSpin is able to also give better categorization accuracy than the method in [14] although it uses significantly less information.

5.2 *Category Prediction*

In this section, we consider the problem of *category prediction* (object recognition). While this is a different problem from the main focus of this paper: topic discovery, we will show that our framework can be also used to perform category prediction. Moreover, we demonstrate that our approach yields superior prediction accuracy results to the state of the art [14].

Each image in the BMW dataset can be represented by a triple (a, b, c), where c is the category number (0–19), b is camera number (0–4) and a is the shot number (0–15). Let M consist of all images with odd a and $b = 2$ (i.e., 8 images per category). The remaining images are partitioned into two groups: T consisting of images with even a and $b \neq 2$ (32 images per category) and D consisting of the rest of the images (40 images per category). Finally, let L be the set of all images with $b = 2$. We will set aside L for "learning", M for "matching" and T for "testing" as described below.

In the following, we will describe and compare four methods, two from the literature (Baseline and NYS [14]) and two new ones (Method 1 and Method 2). All of the methods perform the following category prediction task: Using images in L, learn a classifier which matches each image i in the testing set T to an image $m(i)$ in the matching set M.

1. *Baseline.* This classifier is given by $m(i) = \arg\min\{\|h^i - h^m\|_1 : m \in M\}$. That is, we assign i to image $m(i)$ whose histogram is closest to that of i in L_1 norm.
2. *NYS.* In [14], the authors for each category c form a matrix A_c of normalized histograms corresponding to images in L having category c, and then extract several sparse PCs of A_c. Let the union of the supports of the PCs for every c be I_c, and let $I = \cup_c I_c$. The NYS classifier is given by $m(i) = \arg\min\{\sum_{j \in I} |h^i_j - h^m_j| : m \in M\}$. This is similar to baseline, with the difference that only the important features (I) are used when computing the L_1 distance.
3. *Method 1.* Here, we propose a classifier similar to NYS with the exception that I is obtained as the union of the supports of 160 50-sparse PCs of matrix A_L whose rows are the normalized histograms of all images in L.
4. *Method 2.(TopSpin)* Here, we compute 160 50-sparse PCs from A_L and assign each PC to the image in M with which it has the highest interference, where interference is measured as in Eq. (2). Then, when querying an image from T, we assign it to the PC with which it has the highest interference and through this, using the mapping just described, to an image in M.

For all methods, a dictionary of p = 5,000 visual words is first extracted from images in L, and then normalized histograms are computed for all images. We use the AM-SPCA algorithm discussed in detail in Sect. 4 for PC extraction in all SPCA based methods, i.e., all methods but the baseline. We compute the prediction accuracy of each method defined as the percentage of images $i \in T$ for which i and $m(i)$ have the same category. The results are summarized in Table 1.

We observe from Table 1 that Method 2 is best then follows Method 1, which is in turn superior to both NYS and Baseline. Method 2 outperforms Baseline by cca 4%. This experiment suggests that interference works much better than using L_1-norm. We also observe that computing PCs using all of A_L is better than computing PCs separately for each class, which is especially encouraging since computing the PCs for all training images does not need category information apriori. That is, Method 1 and 2 which are unsupervised learning methods perform better than NYS, which is a supervised learning method.

Table 1 Category prediction accuracy of four methods; by category (only first 10 categories shown) and total. Our approach improves on Baseline by 4%

Cat.	Baseline (%)	NYS (%)	Method 1 (%)	Method 2 (%)
0	100.00	100.00	100.00	100.00
1	90.62	93.75	90.62	87.50
2	68.75	71.88	68.75	87.50
3	96.88	96.88	100.00	96.89
4	81.25	81.25	81.25	100.00
5	100.00	100.00	100.00	100.00
6	100.00	100.00	100.00	81.25
7	81.25	81.25	84.38	96.88
8	37.50	43.75	37.50	81.25
9	40.62	46.88	46.88	75.00
10–19	...	...	...	...
Total (%)	84.53	84.68	85.16	**88.44**

Remark 1 Note that T consists precisely of those images which do not have neither a nor b in common with any images in M. This is crucial as for every image in D, there is a (very) similar image in M (one with the same a but taken by a different camera b), which may skew the results. In fact, prediction accuracy on images from group T is 69.2187% and on images from group D is 89.8750%, if we choose SURF descriptors and $p = 1{,}000$, as in [14]. This gap is present also when SIFT and $p = 5{,}000$ is used, where the accuracy for group T is 84.5313% and for group D is 96.6250%. This is the reason why we have discarded D and used only T for testing.

Remark 2 Also note that a perfect comparison of our results with [14] is not possible as all the data needed to reproduce the experiments exactly as in [14] is not available to us. After implementing their method and setting all available options, we obtained Baseline prediction accuracy 80.69%, whereas in [14] the reported figure is 80.02%.

Let us now look at the features (visual words) selected by the four methods described above. In Fig. 6, we show 3 images from different categories. The first row shows all features in the dictionary appearing on these images. These are the features used by Baseline. The second row shows only features in I_c, for the three different values of category c the three images belong to. In the third row, we show the aggregate features $I = \cup_c I_c$. Finally, the last row shows the features selected our approach (Method 1/Method 2). Note that we are able to achieve a better selection of features than NYS (third row) *without* the knowledge of image categories. The number of selected features for both NYS, Method 1, and Method 2 was chosen to be the same for fairness of comparison purposes.

In Fig. 7, we give an additional insight into why Method 2 (i.e., TopSpin) work. The horizontal axis represents all test images belonging to three categories, CAT1, CAT2 and CAT3. These are the categories that the three images in Fig. 6 belong to. That is,

Fig. 6 Features (visual words) selected by Baseline (first row), NYS per category = I_c (second row), NYS in aggregate = I (third row), and Method 1/Method 2 (last row)

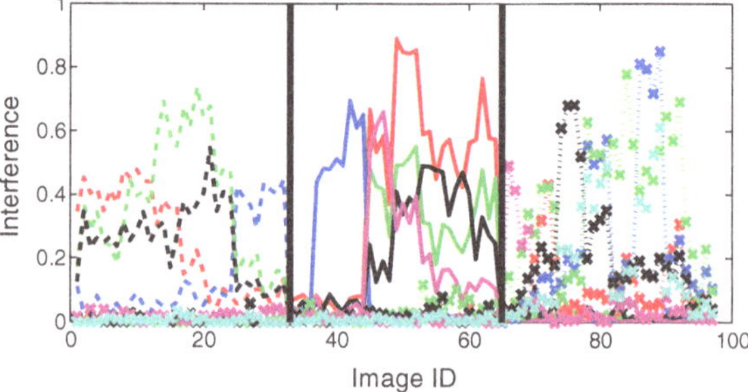

Fig. 7 160 PCs represented as 160 lines with unique formatting, and their interference with 96 images (32 test images from 3 categories). We see that each PC has high interference with a subset of images of a single category only, effectively selecting it

we consider 32 × 3 images. The first 32 images correspond to CAT1; images 33–64 to CAT2 and images 65–96 belong to CAT3. Now, for *all* 160 sparse PCs we plot a unique line in this plot, representing the interference of that PC with the images. For instance, the PC represented by the solid red line has high interference with images 45–64. Notice that all these images belong to CAT2. The PC corresponding to the solid blue line has high interference with 34–45, again a subset of images of CAT2. Note that neither the solid blue nor the solid red line have peaks in any of the other two regions/categories. This means that the PCs representing them effectively represent some object common to a subset of images in CAT2. The same is true for all other lines and the PCs they represent.

5.3 Comparison with LDA

Latent Dirichlet Allocation (LDA) is a probabilistic generative model of a corpus (corpus: a collection of images, documents, or other objects depending on the application). It assumes that there are hidden (latent) variables that correspond to a number of themes (or topics), and aims to infer this hidden structure using observed variables, e.g., the visual words, via an *(approximate) posterior inference* method such as a mean field variational method [4] or Gibbs sampling. In our context, the basic LDA model assumes that each image is a mixture of hidden topics. Each topic corresponds to a probability distribution (over the set of all visual words or the dictionary) and each visual word is drawn from one of these distributions. The distribution of topic distributions as well as the distributions themselves are assumed to be Dirichlet with a set of parameters, that play a significant role on the sparsity, mean, and skewness of the topic distributions and the proportion of topics assigned to each document. LDA has been quite successful in topic discovery and has been modified and extended by various researchers such as the Supervised LDA ([3]).

In this section, we compare our new algorithm with the LDA algorithm.[3] Both algorithms are designed to perform the same task of discovering hidden topics. However, there are few challenges in designing an experiment to make such a comparison in a totally fair fashion. First, it is assumed that LDA has prior information on the number of hidden topics in the document. In our case, even if we have such information, this does not necessarily give the number of sparse PCs that need to be extracted to discover all the topics. It is common for TopSpin to identify two or more PCs that correspond to one topic, each capturing a *subtopic*. This is a disadvantage on one hand since there is some uncertainty in the number of PCs to be extracted, but it also has a certain advantage in the sense that using TopSpin, one can discover hidden topics in more detail and discover new topics. Also, TopSpin is more robust to the number of hidden topics. Consider a scenario, where LDA has been used to identify 100 hidden topics. Nevertheless, after the experiment is completed, new information is obtained that the number of topics is actually much larger, say 150. In this case,

---

[3]We use http://www.cs.princeton.edu/~blei/lda-c/ implementation for LDA.

to discover 150 topics correctly, the LDA algorithm needs to be run from scratch. On the other hand, the topic discovery in TopSpin is *sequential* and *incremental*, i.e., one can continue to extract more sparse PCs after an experiment is completed if it is decided that the number of topics discovered is not enough.

The Dataset: We performed all the experiments on Caltech 101 datasets as in [7], which has 8,677 images in 101 different categories.

Now we will describe the measures we have used to compare LDA and TopSpin methods.

Purity and Conditional Entropy Given ground truth category labels X and estimated cluster labels Y, *purity* is defined in [19] as the mean of the maximum class probabilities, i.e.,

$$Purity(X|Y) = \sum_{y \in Y} p(y) \max_{x \in X} p(x|y). \tag{4}$$

The distribution $p(x, y)$ is estimated from the observed frequencies in a dataset (and hence we can compute only empirical purity). Note that, purity is always in [0, 1] and a higher purity measure is desirable.

Another frequently used metric is the *mutual information* [19] or the gain in entropy:

$$I(X|Y) = H(X) - H(X|Y), \tag{5}$$

where

$$H(X|Y) = \sum_{y \in Y} p(y) \sum_{x \in X} p(x|y) log \frac{1}{p(x|y)}. \tag{6}$$

As the base entropy $H(X)$ in our case is constant, we can use the conditional entropy $H(X|Y)$ in lieu of mutual information for direct comparison of different algorithms. Intuitively, conditional entropy (CE) measures how much uncertainly remains in the true category labels X given the estimated cluster labels Y. Thus, a smaller CE is desirable.

The comparisons of LDA and TopSpin w.r.t. purity and CE are shown in Table 2 (along with other metrics to be discussed in more details). Two different dictionaries of size 10 and 100 K were used. The configuration of TopSpin used in the experiments is also shown in the table caption. The same configuration was used for both dictionaries and all metrics. We have set the number of topics in LDA to the correct level (101). For TopSpin, we extracted 101 sparse PCs using the AM algorithm. The results show TopSpin performs better than LDA consistently across all the metrics.

We further provide results of different TopSpin configurations. We examined 4 popular algorithms for extracting the features (MSER-SURF, MSER-SIFT, SIFT-SIFT, and SURF-SURF methods). We tried the following levels of sparsity {5, 10, 20, 40, 80, 160, 320, 640, 1280} and two versions of weights (tf-idf and equal weights). We report the TopSpin results with the best purity measure and corresponding CE among the various parameters we tried in Table 3. In Table 4, we provide more information, in particular, compare the two options for the weights (tf-idf and constant

Table 2 Summary of comparison of LDA and TopSpin with respect to Purity, CE, PA, PAR, CD. TopSpin configuration: SURF-SURF, level of sparsity 20, tf-idf weights, at every image we assumed only words which appears at least 4 times

Dic.	Metric	LDA	TopSpin
10k	Purity	0.2890	**0.3536**
10k	CE	2.7068	**0.5738**
10k	PA	7.0196	**13.9118**
10k	PAR	0.0277	**0.0846**
10k	CD	**48**	**48**
100k	Purity	0.2543	**0.3939**
100k	CE	2.8895	**0.4554**
100k	PA	2.7059	**19.9902**
100k	PAR	0.0114	**0.1065**
100k	CD	32	**55**

Table 3 Comparison of LDA and TopSpin with respect to Purity and Conditional Entropy

Dic.	Method	Purity		Con. Entropy	
		LDA	TopSpin	LDA	TopSpin
10k	MSER-SURF	0.19534	**0.5411**	3.23012	**0.9692**
10k	MSER-SIFT	0.36610	0.3052	2.38435	2.8661
10k	SIFT-SIFT	0.25353	**0.2761**	2.93530	**0.8112**
10k	SURF-SURF	0.28904	0.2126	2.70675	3.4459
100k	MSER-SURF	0.14694	**0.5360**	3.47104	**0.9815**
100k	MSER-SIFT	0.30852	**0.3515**	2.77324	**2.6525**
100k	SIFT-SIFT	0.19764	**0.3905**	3.24172	**0.7593**
100k	SURF-SURF	0.25427	**0.3477**	2.88951	**0.3546**

Table 4 Comparison of LDA and two versions of TopSpin with respect to Purity

Dic.	Method	LDA	TopSpin/tf-idf	TopSpin/const.
10k	MSER-SURF	0.19534	**0.5411** (5)	**0.4739** (20)
10k	MSER-SIFT	0.36610	0.2912 (160)	0.3052 (160)
10k	SIFT-SIFT	0.25353	**0.2761** (5)	0.2145 (80)
10k	SURF-SURF	0.28904	0.2119 (10)	0.2126 (20)
100k	MSER-SURF	0.14694	**0.5360** (80)	**0.5357** (5)
100k	MSER-SIFT	0.30852	**0.3515** (640)	**0.3325** (1280)
100k	SIFT-SIFT	0.19764	**0.3905** (5)	**0.3783** (5)
100k	SURF-SURF	0.25427	**0.3477** (5)	**0.3074** (5)

weights) separately and also report the sparsity level in parentheses. We observe that on average over the 8 experiments we have conducted, TopSpin performs better than the LDA algorithm. When the dictionary is large, then TopSpin outperforms LDA regardless of the method used for extracting the visual words or the weights that are used in the interference calculation. It is interesting to observe that very sparse PCs work quite well in general and MSER-SIFT method seems to require denser PCs to work well.

Class Diversity and Prediction Accuracy Given a dataset consisting of T different classes one would like to discover, ideally, k hidden topics in such a way that every hidden topic capture different ground truth category. However, more likely, the k hidden topics that have been discovered by the algorithm (LDA or TopSpin) will cover only a subset of truth classes. We define and use new measures that are related to the accuracy of the category prediction using number of categories that are correctly discovered by a topic discovery algorithm.

Given a hidden topic (either a probability distribution as in LDA or a sparse PC in TopSpin), we sort (in descending order) the images with respect to the corresponding probabilities (that a given image belongs to the given hidden topic for LDA) or the interferences (between the given image and the given sparse PC for TopSpin). We define *prediction accuracy*, denoted as PA, as follows:

$$\mathrm{PA} := \frac{1}{k} \sum_{j=1}^{k} C_j,$$

where C_j is a biggest number such that the first C_j images (with respect to the jth hidden topic or sparse PC as described above) belong to the same ground truth class and k is the number of hidden topics used in discovery or the number of sparse PCs extracted.

We also have a slightly modified measure which capture the fact that ground truth classes can have different cardinalities. The modified measure is referred to as the *relative prediction accuracy*, denoted as PAR, and defined as

$$\mathrm{PAR} := \frac{1}{k} \sum_{j=1}^{k} \frac{C_j}{\mathcal{C}_j},$$

where $\mathcal{C}_j$ is the cardinality of ground truth class to which the first C_j images (corresponding to the jth hidden topic or sparse PC) belong to.

The PA and PAR measure are related to the accuracy of our prediction in terms of categorizing images together correctly with respect to the given ground truth categories. In addition to this, we are also interested in the number of categories that are discovered by the topic discovery algorithms. This is due to the fact that, even when we set the number of hidden topics or the sparse PCs to the number of categories in the test data, it is not certain that each topic or PC will correspond to a different category. We define the *class diversity* measure, denoted as CD, to assess the diversity of classes discovered as follows:

$$\text{CD} := \#\text{of ground truth classes captured by } k \text{ hidden topics}$$

Table 2 compares LDA and TopSpin in terms of these three new measures. The same TopSpin configuration as in Purity/CE comparisons was used here. The results show consistently improved performance using TopSpin.

We further examine different TopSpin configurations w.r.t. PA, PAR, and CD. Results are shown in Table 5. As before, we try two different size of dictionaries, four different methods for feature extraction, two strategies for assigning weights, and various levels of sparsity. We report the best TopSpin results in terms of sparsity level of PCs. Clearly, the performance of LDA is getting worse with increasing size of dictionary, whereas the opposite is true for TopSpin. For the larger dictionary,

Table 5 Comparison of LDA and TopSpin with respect to class diversity and (relative) prediction accuracy

Dic.	Method	Weights	LDA			TopSpin			
			CD	PA	PAR	Spars.	CD	PA	PAR
10k	MSER-SURF	tf-idf	39	1.3333	0.0117	5	**43**	**24.0196**	**0.1147**
		const.				20	39	**18.2843**	**0.0765**
10k	MSER-SIFT	tf-idf	45	8.4804	0.0278	160	37	**9.5000**	**0.0383**
		const.				80	30	**13.5392**	**0.0642**
10k	SIFT-SIFT	tf-idf	42	2.6176	0.0131	5	**44**	**7.4020**	**0.0495**
		const.				5	38	**7.6961**	**0.0467**
10k	SURF-SURF	tf-idf	48	7.0196	0.0277	5	26	3.7059	0.0126
		const.				40	16	5.4706	0.0138
100k	MSER-SURF	tf-idf	18	1.9020	0.0070	640	**37**	**25.8627**	**0.1001**
		const.				640	**37**	**25.9510**	**0.0992**
100k	MSER-SIFT	tf-idf	21	1.8235	0.0081	640	**38**	**24.5196**	**0.0798**
		const.				320	**22**	**24.7549**	**0.0621**
100k	SIFT-SIFT	tf-idf	24	1.8627	0.0095	10	**45**	**15.8922**	**0.0977**
		const.				5	**44**	**15.5588**	**0.0969**
100k	SURF-SURF	tf-idf	32	2.7059	0.0114	5	**38**	**18.3039**	**0.0993**
		const.				5	**42**	**13.7157**	**0.0884**

TopSpin outperforms LDA significantly. Although, we have reported the best results with respect to the sparsity level, we like to note that TopSpin outperforms LDA for most of the sparsity levels we have tried. It seems that the conclusion is not clear for the smaller dictionary size. LDA seems to discover more categories than TopSpin (on the average), but the precision accuracy is usually much higher for TopSpin. In

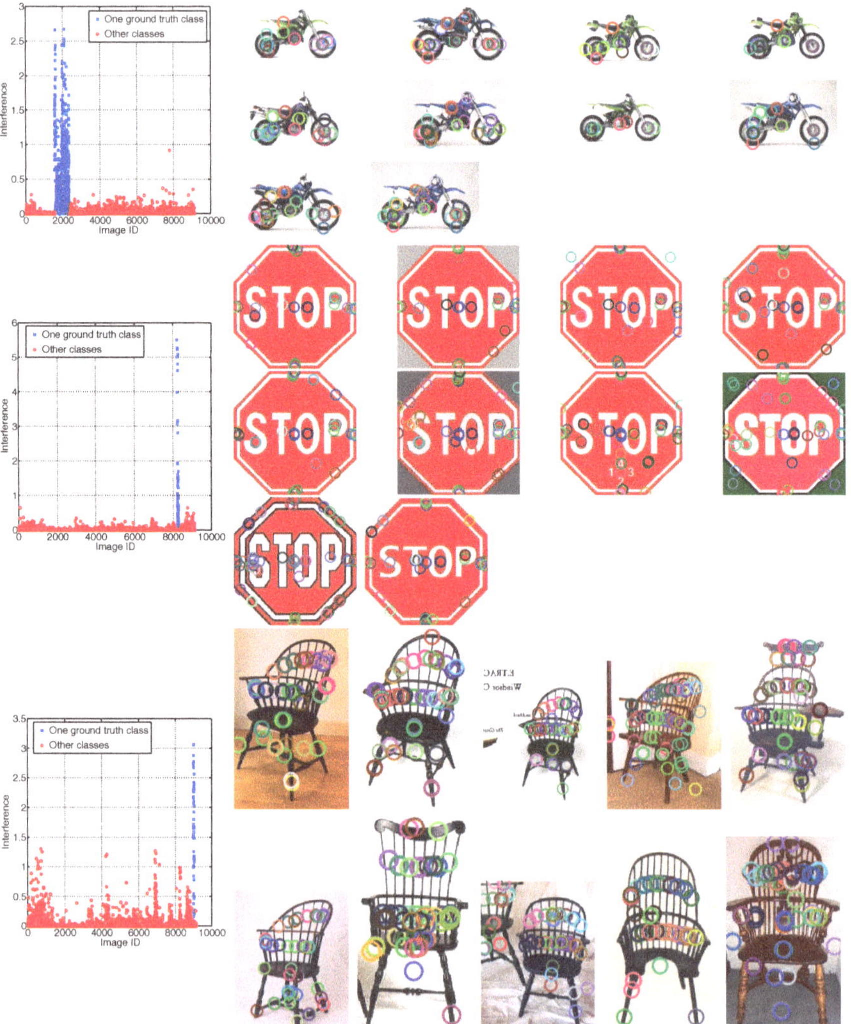

Fig. 8 Example of images with highest interference for three sparse PCs. In each row, we first plot the interferences for all images in the dataset (left) with one sparse PC and then provide 10 images with the highest interference with this PC on the right

addition, we also observe that tf-idf weights usually perform better than the constant weights for both dictionary sizes.

Example of TopSpin's Output We also provide a sample of the output obtained by the TopSpin algorithm in Fig. 8, where we have used three sparse components. On the left column, we plot the interferences of images with the three sparse PCs (one in each row). We observe that there are few images which have high interference with each PC. We then provide the 10 images whose interference with each sparse PC is the highest on the right column. We see that the first PC discovered the motorbike, second PC discovered the STOP sign, and the third PC discovered the chair images from thousands of images. We also marked the ground truth class of motorbike, STOP sign, and Chair on the left plots with blue squares. It can be seen that the interference of all the images in these classes with the corresponding sparse PCs is very high.

6 Contributions

We now summarize some of our main contributions:

1. We have developed an algorithm (TopSpin) for solving the problem: topic discovery in a collection of unlabeled images. Our algorithm applies Sparse PCA to identify co-occurred visual words that can be used as topic signatures.
2. We have demonstrated on real datasets that TopSpin is able to discover topics and correctly assign images to the topics.
3. When used for category prediction, our framework gives higher accuracy than that of [14]. Moreover, this is achieved without knowing what the categories are as sparse PCA is applied to data coming from all (test) images of all categories, not to (test) images of each category individually as in [14].
4. Our Sparse PCA solver is 3 or more order of magnitude faster than ALM. It solves the Sparse PCA problem directly (i.e., not a relaxation), and unlike ALM, has direct control over the sparsity of the PCs (via s). Our Sparse PCA solver is parallel in nature and scalable to high dimensions.
5. Using a large corpus, we show that the TopSpin algorithm outperforms the LDA algorithm (especially for large dictionary size) in terms if topic and category discovery. t discovers the categories of images more precisely and discovers more hidden topics correctly. Since LDA is one of the most popular and successful methods developed for this purpose, this is a remarkable success for the TopSpin algorithm.

References

1. Bart, E., Porteous, I., Perona, P., Welling, M.: Unsupervised learning of visual taxonomies. In: CVPR (2008)
2. Blei, D.M., Griffiths, T.L., Jordan, M.I., Tenenbaum, J.B.: Hierarchical topic models and the nested Chinese restaurant process. In: NIPS (2004)
3. Blei, D.M., McAuliffe, J.: Supervised topic models. In: NIPS (2007)
4. Blei, D.M., Ng, A.Y., Jordan, M.I.: Latent dirichlet allocation. J. Mach. Learn. Res. **3**, 993–1022 (2003). Mar
5. d'Aspremont, A., Bach, F., Ghaoui, L.E.: Optimal solutions for sparse principal component analysis. J. Mach. Learn. Res. **9**, 1269–1294 (2008)
6. d'Aspremont, A., Ghaoui, L.E., Jordan, M.I., Lanckriet, G.R.G.: A direct formulation for sparse PCA using semidefinite programming. SIAM Rev. **48**(3), 434–448 (2007)
7. Fei-Fei, L., Fergus, R., Perona, P.: Learning generative visual models from few training examples: an incremental bayesian approach tested on 101 object categories (2004)
8. Grauman, K., Darrell, T.: Unsupervised learning of categories from sets of partially matching image features. In: CVPR (2006)
9. Journée, M., Nesterov, Y., Richtárik, P., Sepulchre, R.: Generalized power method for sparse principal component analysis. J. Mach. Learn. Res. **11**, 517–553 (2010)
10. Kinnunen, T., Kamarainen, J.-K., Lensu, L., Kalviainen, H.: Unsupervised visual object categorisation via self-organisation. In: ICPR (2010)
11. Lowe, D.: Object recognition from local scale-invariant features. In: ICCV (1999)
12. Mackey, L.: Deflation methods for sparse PCA. In: NIPS (2008)
13. Naikal, N., Yang, A., Sastry, S.: Towards an efficient distributed object recognition system in wireless smart camera networks. In: International Conference on Information Fusion (2010)
14. Naikal, N., Yang, A.Y., Shankar Sastry, S.: Informative feature selection for object recognition via sparse PCA. In: ICCV (2011)
15. Nister, D., Stewenius, H.: Scalable recognition with a vocabulary tree. In CVPR (2006)
16. Richtárik, P., Takáč, M., Ahipasaoglu S.D.: Alternating maximization: unifying framework for 8 sparse PCA formulations and efficient parallel codes (2012). arXiv:1212.4137
17. Sivic, J., Russell, B.C., Zisserman, A., Freeman, W.T., Efros, A.A.: Unsupervised discovery of visual object class hierarchies. In: CVPR (2008)
18. Sivic, J., Zisserman, A.: Video google: a text retrieval approach to object matching in videos. In: ICCV (2003)
19. Tuytelaars, T., Lampert, C.H., Blaschko, M.B., Buntine, W.: Unsupervised object discovery: a comparison. IJCV **88**(2) (2010)
20. J. Zhang, M. Marszalek, S. Lazebnik, and C. Schmid. Local features and kernels for classification of texture and object categories: a comprehensive study. IJCV (2007)
21. Zhang, Y., Ghaoui, L.E.: Large–scale sparse principal component analysis with application to text data. In: NIPS (2011)
22. Zou, H., Hastie, T., Tibshirani, R.: Sparse principal component analysis. Technical report, Stanford University (2004)

Exact Optimal Solution to Nonseparable Concave Quadratic Integer Programming Problems

Fenlan Wang

Abstract Nonseparable quadratic integer programming problems have extensive applications in real world and have received considerable attentions. In this paper, a new exact algorithm is presented for nonseparable concave quadratic integer programming problems. This algorithm is of a branch and bound frame, where the lower bound is obtained by solving a quadratic convex programming problem and the branches are partitioned via a special domain cut technique by which the optimality gap is reduced gradually. The optimal solution to the primal problem can be found in a finite number of iterations. Numerical results are also reported to illustrate the efficiency of our algorithm.

Keywords Nonseparable concave integer programming problems · Linear and contour cut · Domain partition · Quadratic convex programming

MSC (2010): 90C10 · 90C26 · 90C30

1 Introduction

In this paper, the following nonseparable concave integer programming problems are considered:

$$
\begin{aligned}
(P) \qquad & \min f(x) = \frac{1}{2}x^T Qx + c^T x \\
& \text{s.t. } Ax \le b, \\
& \quad x \in X = \{x \in \mathbb{Z}^n \mid l \le x \le u\},
\end{aligned}
$$

F. Wang
College of Science, Nanjing University of Aeronautics and Astronautics, Nanjing, China
e-mail: flwang@nuaa.edu.cn

J. D. Pintér and T. Terlaky (eds.), *Modeling and Optimization: Theory and Applications*, Springer Proceedings in Mathematics & Statistics 279,
https://doi.org/10.1007/978-3-030-12119-8_9

where $Q \in \mathbb{R}^{n\times n}$ is a negative definite symmetric matrix, $c \in \mathbb{R}^n, A \in \mathbb{R}^{m\times n}, b \in \mathbb{R}^m$, and $\mathbb{Z}^n$ is the set of all integer points in $\mathbb{R}^n$. Further, $l = (l_1, l_2, \ldots, l_n)^T$ and $u = (u_1, u_2, \ldots, u_n)^T \in \mathbb{Z}^n$ are the lower and upper bounds of the variable x, respectively.

Nonlinear integer programming models have many applications in real world, where convex and concave quadratic cost functions are often encountered, such as capital budgeting [17], capacity planning [4], optimization problems from graph theory [2, 13], fixed charge problems with integer variables [10], and problems involving economies of scale. There are many methods in the literature for nonlinear separable integer programming problems. Ibaraki and Katoh [11] summarized certain algorithms for single constrained resource allocation problems, where the single constraint is of a special form of $\sum_{j=1}^{n} x_j = N$. A hybrid approach was proposed in [16] for separable nonlinear integer programming problems with a nonincreasing objective function. Hochbaum studied a single constrained problem, where both the objective function and the constraint function are convex and monotonically nonincreasing in [9]. Betthauer and Shetty [6] proposed a branch-and-bound algorithm for single constrained separable integer programming problems where the objective function and the constraint function are all convex functions. A pegging algorithm was presented in [5] for nonlinear resource allocation problems. Most of the existing methods for separable nonlinear integer programming problems are dynamic programming-based methods and continuous relaxation-based branch and bound methods, see [3, 7, 12, 20, 21].

For nonseparable integer programming problems, Lagrangian decomposition methods were presented in [8, 18, 19, 23] for nonseparable convex knapsack problems. These methods overcome the difficulty caused by the nonseparability. However, there are only a few implemented methods for nonseparable concave knapsack problems.

In addition, a few novel domain cut techniques were presented in [14, 15, 22] for separable integer programming problems. These methods inspired us to develop the algorithm in this paper for nonseparable concave integer programming problems.

In this paper, a new exact method is presented for problem (P). The proposed algorithm is essentially a branch and bound method, where the lower bound is obtained by solving a quadratic convex programming problem, and the branch is partitioned via a linear and contour cut technique. The lower bound can be used to discard some integer sub-boxes that do not contain feasible solutions better than the incumbent one, and the domain cut technique helps us to reduce the optimality gap. The presented algorithm can find an optimal solution to the primal problem in a finite number of iterations.

The paper is organized as follows. In Sect. 2, a lower bound for each subproblem is derived. Section 3 gives a domain cut and partition technique. The algorithm is formally described in Sect. 4 and is illustrated by a small numerical example. Finally, computational results for randomly generated nonseparable concave knapsack problems are preliminarily reported in Sect. 5.

2 A Lower Bound for the Subproblems

Let $\alpha, \beta \in \mathbb{Z}^n$. Denote by $[\alpha, \beta]$ the box (hyper-rectangle) formed by α and β, $[\alpha, \beta] = \{x \mid \alpha_j \le x_j \le \beta_j, j = 1, \ldots, n\}$. Denote by $\langle \alpha, \beta \rangle$ the set of integer points in $[\alpha, \beta]$,

$$\langle \alpha, \beta \rangle = \{x \mid \alpha_j \le x_j \le \beta_j,\ x_j \text{ integer},\ j = 1, \ldots, n\} = \bigotimes_{j=1}^{n} \langle \alpha_j, \beta_j \rangle.$$

Define $[\alpha, \beta] = \langle \alpha, \beta \rangle = \emptyset$ if $\alpha \not\le \beta$. Let $\langle \alpha, \beta \rangle \subseteq X$ be nonempty and (SP) be a subproblem of (P) by replacing X by $\langle \alpha, \beta \rangle$ with $\alpha \le \beta$. For convenience, the set $\langle \alpha, \beta \rangle$ is called an integer box.

Now consider the subproblem of (P):

$$\begin{aligned} (SP) \qquad & \min f(x) = \frac{1}{2}x^T Q x + c^T x \\ & \text{s.t. } Ax \le b, \\ & \quad x \in \langle \alpha, \beta \rangle. \end{aligned}$$

We know the objective function can be always rewritten as the following equivalent form:

$$\begin{aligned} f(x) &= \frac{1}{2}x^T Q x + c^T x + \frac{p}{2}\sum_{i=1}^{n} x_i^2 - \frac{p}{2}\sum_{i=1}^{n} x_i^2 \\ &= f_1(x) + f_2(x), \end{aligned}$$

where p is a positive parameter, $f_1(x) = \frac{1}{2}x^t Q x + c^T x + \frac{p}{2}\sum_{i=1}^{n} x_i^2$ and $f_2(x) = -\frac{p}{2}\sum_{i=1}^{n} x_i^2$. Note that $f_1(x)$ is a convex function if we take p greater than the absolute value of the most negative eigenvalue of Q and $f_2(x)$ is a separable concave function.

Now substituting $f_2(x)$ with its linear underestimation function and discarding the integrality condition, we have the following convex quadratic programming problem:

$$\begin{aligned} (P') \quad \min \quad & \varphi(x) = \frac{1}{2}x^T Q x + c^T x + \frac{p}{2}\sum_{i=1}^{n} x_i^2 + \frac{p}{2}\sum_{i=1}^{n} \phi_i(x_i) \\ \text{s.t. } & Ax \le b \\ & x \in [\alpha, \beta] \end{aligned}$$

where

$$\phi_i(x_i) = \begin{cases} -(\beta_i + \alpha_i)x_i + \beta_i\alpha_i, & \alpha_i < \beta_i, \\ -(\alpha_i)^2, & \alpha_i = \beta_i. \end{cases}$$

Let $S = \{x \in [\alpha, \beta] \mid Ax \leq b\}$. Obviously, $\varphi(x) \leq f(x)$ for $\forall x \in S$. Further, let $v(\cdot)$ denote the optimal value of problem $(\cdot)$. Thus, we can obtain easily a lower bound $v(P')$ for the optimal objective function value of (SP) by solving the quadratic convex relaxation problem (P').

3 Linear and Contour Cuts

By solving problem (P'), we can obtain a continuous optimal solution $\hat{x}$ of (P') and then $v(P')$ is a lower bound of $v(SP)$. If $\hat{x}$ is an integer solution, then $\hat{x}$ is also a feasible solution to (SP) and $v(P')$ is an upper bound of $v(P)$. If $\hat{x}$ is not an integer solution, then we can obtain two integer points x^1 and x^2 by rounding $\hat{x}$ up or down.

If $\nabla\varphi(\hat{x}) \neq 0$, then we have the following lemma.

Lemma 3.1 *Let $\hat{x}$ be a continuous optimal solution to (P'). Let x^1 and x^2 be two integer points by rounding $\hat{x}$ up and down along two directions d_1 and d_2, respectively, where $d_1^T \nabla\varphi(\hat{x}) > 0$ and $d_2^T \nabla\varphi(\hat{x}) < 0$. Then*

(i) x^2 must be an infeasible solution and there must be no feasible solutions in the set $\nabla\varphi(x^2)^T(x - x^2) < 0$.

(ii) x^1 may be feasible or infeasible. If x^1 is an infeasible solution, without loss of generality, suppose $A_i x^1 > b_i$, where $A_i = (a_{i1}, a_{i2}, \ldots, a_{in})$. Then, the set $A_i(x - x^1) \geq 0$ does not contain any feasible solution.

Proof (i) Suppose x^2 is feasible. Then $\varphi(x^2) < \varphi(\hat{x})$ holds, since x^2 is obtained along the direction d_2 which is a decreasing direction. Thus, x^2 has a better objective value than the optimal solution of problem (P'), which is contradiction to that $\hat{x}$ is an optimal solution to (P').

For $\forall x \in \{x \mid \nabla\varphi(x^2)^T(x - x^2) < 0\}$, the vector $x - x^2$ is a decreasing direction and analogous to the previous argument. x is also infeasible.

(ii) Since $A_i x^1 > b_i$ and $A_i(x - x^1) \geq 0$, then $A_i x \geq A_i x^1 > b_i$ must hold. This implies the set $A_i(x - x^1) \geq 0$ does not contain any feasible solution. □

Based on the above lemma, the following domain cuts are derived to cut certain integer sub-boxes from $\langle\alpha, \beta\rangle$ to reduce the optimality gap. For convenience, denote by $N_1(\bullet)$, $N_2(\bullet)$, and $F(\bullet)$, the following three different integer boxes that will be cut off.

By statement (i) of Lemma 3.1, there is no feasible solutions in the integer box $N_1(x^2) = \langle\gamma, \delta\rangle$, where γ_j, δ_j are determined by

$$\gamma_j = \begin{cases} (x^2)_j, & (\nabla\varphi(x^2))_j < 0, \\ \alpha_j, & (\nabla\varphi(x^2))_j > 0, \end{cases} \tag{3.1}$$

$$\delta_j = \begin{cases} \beta_j, & (\nabla\varphi(x^2))_j < 0, \\ (x^2)_j, & (\nabla\varphi(x^2))_j > 0. \end{cases}$$

By statement (ii) of Lemma 3.1, if x^1 is infeasible, let us suppose that $A_i x^1 > b_i$, where $A_i = (a_{i1}, a_{i2}, \ldots, a_{in})$. We can cut the integer box $N_2(x^1) = \langle \gamma, \delta \rangle$ without missing any feasible solution to (SP), where

$$\gamma_j = \begin{cases} (x^1)_j, & (A_i)_j > 0, \\ \alpha_j, & (A_i)_j < 0, \end{cases} \tag{3.2}$$

$$\delta_j = \begin{cases} \beta_j, & (A_i)_j > 0, \\ (x^1)_j, & (A_i)_j < 0. \end{cases}$$

For a feasible solution, e.g., if x^1 is a feasible solution, let us consider the following ellipsoid contour:

$$\frac{1}{2}x^T Qx + c^T x = v, \tag{3.3}$$

where $v = f(x^1)$. Since Q is a negative definite symmetric matrix, there exists an orthogonal matrix P, such that $P^{-1} = P^T$ and $P^T QP = \Lambda = Diag(\lambda_1, \lambda_2, \ldots, \lambda_n)$ with all $\lambda_i < 0,\ i = 1, 2, \ldots, n$. Let $x = Py$, then

$$\frac{1}{2}x^T Qx + c^T x = \frac{1}{2}y^T \Lambda y + c^T Py = \frac{1}{2}\sum_{i=1}^{n}(\lambda_i y_i^2 + 2s_i y_i) = \frac{1}{2}\sum_{i=1}^{n}\lambda_i(y_i^2 + 2\frac{s_i}{\lambda_i}y_i).$$

Thus, from (3.3), we have

$$\begin{aligned} v = \frac{1}{2}x^T Qx + c^T x \Leftrightarrow 2v = x^T Qx + 2c^T x &= \sum_{i=1}^{n}\lambda_i(y_i^2 + 2\frac{s_i}{\lambda_i}y_i) \\ &= \sum_{i=1}^{n}[\lambda_i(y_i + \frac{s_i}{\lambda_i})^2 - \frac{s_i^2}{\lambda_i}] \\ \Leftrightarrow \sum_{i=1}^{n}\lambda_i(y_i + \frac{s_i}{\lambda_i})^2 = 2v + \sum_{i=1}^{n}\frac{s_i^2}{\lambda_i} &= 2v + c^T Q^{-1}c, \end{aligned} \tag{3.4}$$

where $c^T P = (s_1, s_2, \ldots, s_n)$. We can easily see that (3.4) is also an ellipsoid in the $y-$space whose center is the point $O' = (-\frac{s_1}{\lambda_1}, \ldots, -\frac{s_n}{\lambda_n})^T$ and the length of the $i-$th axis is $2\sqrt{\frac{2v+c^T Q^{-1}c}{\lambda_i}}$. Therefore, the center of the ellipsoid (3.3) in the $x-$space is the point $O = PO' = -Q^{-1}c$, and the length of its $i-$th axis is also $2\sqrt{\frac{2v+c^T Q^{-1}c}{\lambda_i}}$.

Let $\lceil x \rceil$ and $\lfloor x \rfloor$ be the minimum integer number larger than or equal to x and the maximum integer number less than or equal to x, respectively. The maximum integer box inside the ellipsoid (3.3) is $F(x^1) = \langle \gamma, \delta \rangle$,

$$\gamma = \left(\left\lceil -(Q^{-1}c)_1 - \sqrt{\frac{2v + c^T Q^{-1} c}{n\lambda_{\max}}} \right\rceil, \ldots, \left\lceil -(Q^{-1}c)_n - \sqrt{\frac{2v + c^T Q^{-1} c}{n\lambda_{\max}}} \right\rceil \right), \tag{3.5}$$

$$\delta = \left(\left\lfloor -(Q^{-1}c)_1 + \sqrt{\frac{2v + c^T Q^{-1} c}{n\lambda_{\max}}} \right\rfloor, \ldots, \left\lfloor -(Q^{-1}c)_n + \sqrt{\frac{2v + c^T Q^{-1} c}{n\lambda_{\max}}} \right\rfloor \right),$$

where $\lambda_{\max}$ is the maximum eigenvalue of Q. We can cut the integer box $F(x^1)$ off from $\langle \alpha, \beta \rangle$ without missing any feasible solution better than x^1.

If $\nabla \varphi(\hat{x}) = 0$, then we can obtain two integer points x^1 and x^2 by directly rounding $\hat{x}$ up and down, respectively. Thus, we can also cut the corresponding integer sub-boxes $N_2(x^1)$ or $F(x^1)$, $N_2(x^2)$ or $F(x^2)$ according to x^1, x^2 being feasible or infeasible.

The following lemma will show that $\langle \alpha, \beta \rangle \setminus \langle \gamma, \delta \rangle$ can be partitioned into a union of some integer sub-boxes.

Lemma 3.2 *([14, 15, 22]) Let $\langle \alpha, \beta \rangle$ and $\langle \gamma, \delta \rangle$ are two integer boxes, where α, β, γ, $\delta \in \mathbb{Z}^n$ and $\alpha \le \gamma \le \delta \le \beta$. Then*

$$\begin{aligned} \langle \alpha, \beta \rangle \setminus \langle \gamma, \delta \rangle = &\{\cup_{j=1}^n \left(\Pi_{i=1}^{j-1} \langle \alpha_i, \delta_i \rangle \times \langle \delta_j + 1, \beta_j \rangle \times \Pi_{i=j+1}^n \langle \alpha_i, \beta_i \rangle \right)\} \\ &\cup \{\cup_{j=1}^n \left(\Pi_{i=1}^{j-1} \langle \gamma_i, \delta_i \rangle \times \langle \alpha_j, \gamma_j - 1 \rangle \times \Pi_{i=j+1}^n \langle \alpha_i, \delta_i \rangle \right)\}. \end{aligned} \tag{3.6}$$

In each new generated integer sub-box, the lower bound obtained by solving a quadratic convex programming problem in the previous section and the contour and linear cut technique mentioned in this section can be applied accordingly.

4 The Main Algorithm

This section presents the special domain cut algorithm for solving problem (P).

Algorithm 4.1 (Linear and Contour Cut Algorithm for Nonseparable Concave Integer Programming Problems)

Step 0. (Initialization) Set $f_{\text{opt}} = +\infty$, $X^0 = \{X\}$, $k = 0$.

Step 1. Select the integer sub-box $\langle \alpha^k, \beta^k \rangle$ from X^k that yields the minimum lower bound *LB* obtained by solving the quadratic programming problem (P') in the box $\langle \alpha^k, \beta^k \rangle$. Also, we have a continuous optimal solution $\hat{x}$ to (P').

Step 2. (Domain cut and partition)

- Case (a): $\nabla \varphi(\hat{x}) \ne 0$.
  - If $\hat{x}$ isn't an integer solution, we get two integer points x^1 and x^2 by rounding $\hat{x}$ up or down along two directions d_1 and d_2 respectively, where $d_1^T \nabla \varphi(\hat{x}) > 0$ and $d_2^T \nabla \varphi(\hat{x}) < 0$. Cut off $N_1(x^2)$ defined by (3.1) with α_j, β_j replaced by α_j^k, β_j^k.

If x^1 is a feasible solution, update the incumbent $x_{\text{opt}} := x^1, f_{\text{opt}} := f(x^1)$ if $f(x^1) < f_{\text{opt}}$. Then cut off $F(x^1) \cap \langle \alpha^k, \beta^k \rangle$ from $\langle \alpha^k, \beta^k \rangle$, where $F(x^1)$ is defined by (3.5).
If x^1 is an infeasible solution, cut off $N_2(x^1)$ defined by (3.2) with α_j, β_j replaced by α_j^k, β_j^k from $\langle \alpha^k, \beta^k \rangle$.

- If $\hat{x}$ is an integer solution, update the incumbent $x_{\text{opt}} := \hat{x}$, $f_{\text{opt}} := f(\hat{x})$ if $f(\hat{x}) < f_{\text{opt}}$. Then cut off $N_1(\hat{x})$ and $F(\hat{x})$ from $\langle \alpha^k, \beta^k \rangle$, where $N_1(\hat{x})$, $F(\hat{x})$ are defined by (3.1), (3.5) respectively.

- Case (b): $\nabla \varphi(\hat{x}) = 0$.

Obtain two integer points x^1 and x^2 by directly rounding $\hat{x}$ up and down respectively. Then, cut the corresponding integer sub-boxes $N_2(x^1)$ or $F(x^1)$, $N_2(x^2)$ or $F(x^2)$ according to x^1, x^2 feasible or infeasible.

Then $((\langle \alpha^k, \beta^k \rangle \backslash N_1(x^2)) \backslash N_2(x^1)$ or $((\langle \alpha^k, \beta^k \rangle \backslash N_1(x^2)) \backslash F(x^1)$ is partitioned into a union of some integer sub-boxes by Lemma 3.2, add the generated new integer boxes to the set Y^{k+1}. For each $\langle \alpha, \beta \rangle \in Y^{k+1}$, remove it if it satisfies one of the following situations:

- There are no feasible solutions.
- $LB \geq f_{\text{opt}}$.

Step 3. Set $X^{k+1} = Y^{k+1} \bigcup (X^k \setminus \langle \alpha^k, \beta^k \rangle)$

Step 4. (Termination) If X^{k+1} is empty, STOP and x_{opt} is an optimal solution to (P). Otherwise, set $k := k + 1$, goto Step 1.

Theorem 4.1 *If X is nonempty, then Algorithm 4.1 terminates at an optimal solution of (P) within a finite number of iterations.*

Proof By the linear and contour cut in Step 2, at least x^1 and x^2 are cut off and no optimal solutions are removed in Step 2. Therefore, x_{opt} must be the optimal solution to (P) when the algorithm stops in Step 4 with $X^{k+1} = \emptyset$. The finite termination of the algorithm is obvious by observing the finiteness of X. □

Example 4.1

$$
\begin{aligned}
\min\ & f(x) = \frac{1}{2} x^T Q x + c^T x \\
\text{s.t. } & Ax \leq b, \\
& x \in X = \{x \mid 3 \leq x_i \leq 12,\ x_i \text{ integer},\ i = 1, 2\},
\end{aligned}
$$

where

$$Q = \begin{pmatrix} -2 & 1 \\ 1 & -4 \end{pmatrix}, \; c = \begin{pmatrix} 1 \\ 2 \end{pmatrix}, \; A = \begin{pmatrix} -3 & 2 \\ 1 & 2 \\ 2 & 1 \\ 2 & -6 \\ -4 & -5 \end{pmatrix}, \; b = \begin{pmatrix} -2 \\ 20 \\ 25 \\ -9 \\ -27 \end{pmatrix}$$

The optimal solution is $x^* = (10, 5)^T$ with $f(x^*) = -80$.

Iteration 1

Step 0. Set $f_{\text{opt}} = +\infty$, $X^0 = \{X\}$, $k = 0$.

Step 1. Select X to generate new integer boxes. By solving the quadratic programming problem (P'), we have a lower bound $LB = v(P') = -174.9765$ and a continuous solution $\hat{x} = (6.2774, 6.8613)^T$ in the box X.

Step 2. Obtain two integer points $x^1 = (6, 6)^T$ and $x^2 = (7, 7)^T$ by rounding $\hat{x}$ up and down along two directions d_1 and d_2, respectively. By (3.1), $N_1(x^2) = \langle(7, 7), (12, 12)\rangle$. Also x^1 is feasible and $f(x^1) = -54 < f_{\text{opt}}$, so update the incumbent solution $f_{\text{opt}} := -54$, $x_{\text{opt}} := x^1$. By (3.5), $F(x^1) = \langle(3, 3), (4, 4)\rangle$.

Then $(X \backslash N_1(x^2)) \backslash F(x^1) = \langle(5, 3), (6, 12)\rangle \cup \langle(3, 5), (4, 12)\rangle \cup \langle(7, 3), (12, 6)\rangle$.

For $\langle(5, 3), (6, 12)\rangle$, we have the lower bound $LB = -130.8265$ by solving problem (P') in this box; For $\langle(3, 5), (4, 12)\rangle$, there is no feasible solution in this box, so it is discarded; For $\langle(7, 3), (12, 6)\rangle$, we have the lower bound $LB = -101.6944$ by solving problem (P') in this box.

Step 3. $X^1 = Y^1 = Y_1^1 \cup Y_2^1 = \langle(5, 3), (6, 12)\rangle \cup \langle(7, 3), (12, 6)\rangle$.

Iteration 2

Step 1. Select Y_1^1 with the minimum lower bound to generate new integer boxes. By solving the quadratic programming problem (P') in the box Y_1^1, we also obtain the continuous optimal solution $\hat{x} = (5.5, 7.25)^T$ to (P').

Step 2. Obtain two integer points $x^1 = (5, 7)^T$ and $x^2 = (6, 8)^T$ by rounding $\hat{x}$ up and down along two directions d_1 and d_2, respectively. By (3.1), $N_1(x^2) = \langle(6, 8), (6, 12)\rangle$. And x^1 is infeasible. By (3.2), $N_2(x^1) = \langle(5, 7), (5, 12)\rangle$.

Then $(Y_1^1 \backslash N_1(x^2)) \backslash N_2(x^1) = \langle(5, 3), (5, 6)\rangle \cup \langle(6, 3), (6, 7)\rangle$.

For $\langle(5, 3), (5, 6)\rangle$, we have the lower bound $LB = -50 > f_{\text{opt}}$ by solving problem (P') in this box, so it is discarded; For $\langle(6, 3), (6, 7)\rangle$, we have the lower bound $LB = -72$ and $\hat{x} = (6, 7)^T$ with $f(\hat{x}) = -72$ by solving problem (P') in this box. Since $f(\hat{x}) = -72 < f_{\text{opt}}$, update the incumbent solution $f_{\text{opt}} := -72$, $x_{\text{opt}} := (6, 7)^T$. Now $LB = f_{\text{opt}}$, so this box is discarded.

Step 3. $X^2 = X^1 \backslash Y_1^1 = Y_2^1$.

Iteration 3

Step 1. Select Y_2^1 to generate new integer boxes. By solving the quadratic programming problem (P') in the box Y_2^1, we also have the continuous optimal solution $\hat{x} = (10.0714, 4.8571)$ to (P').

Step 2. Obtain two integer points $x^1 = (10, 4)^T$ and $x^2 = (11, 5)^T$ by rounding $\hat{x}$ up and down along two directions d_1 and d_2, respectively. By (3.1), $N_1(x^2) = \langle(11, 5), (12, 6)\rangle$. And x^1 is also infeasible. By (3.2), $N_2(x^1) = \langle(10, 3), (12, 4)\rangle$.

Then $(Y_2^1 \backslash N_1(x^2)) \backslash N_2(x^1) = \langle(7, 5), (10, 6)\rangle \cup \langle(7, 3), (9, 4)\rangle$.

For $\langle(7, 5), (10, 6)\rangle$, we have the lower bound $LB = -80$ and $\hat{x} = (10, 5)^T$ with $f(\hat{x}) = -80$ by solving problem (P') in this box. Since $f(\hat{x}) = -80 < f_{\text{opt}}$, update the incumbent solution $f_{\text{opt}} := -80$, $x_{\text{opt}} := (10, 5)^T$. Now $LB = f_{\text{opt}}$, so this box is discarded. For $\langle(7, 3), (9, 4)\rangle$, we have the lower bound $LB = -44.7803 > f_{\text{opt}}$ by solving problem (P') in this box, so we discard it.

Step 3. $X^3 = \emptyset$. So $x_{\text{opt}} = (10, 5)^T$ is the optimal solution to (P) with $f_{\text{opt}} = -80$.

5 Computational Results

The algorithm is implemented in FORTRAN 90 and run on a PC with Pentium(R) Dual-core CPU E6700@3.2GHz for problem (P). The algorithm was tested by generating concave quadratic functions $f(x) = \frac{1}{2}x^T Qx + c^T x$, where $Q = (q_{ij}) \in \mathbb{R}^{n\times n}$ and $c \in \mathbb{R}^n$. Ten test problems were randomly generated from uniform distribution for each size n and m. To guarantee that $f(x)$ is a concave function on $[l, u]$, we take $q_{ij} \in [-20, -10]$ $(i \neq j)$, $q_{ii} = \sum_{j\neq i} q_{ij} + d$ with $d \in [-11, -10]$, $c \in [-10, 20]$. The constraint matrix $A = (a_{ij})$ is generated with $a_{ij} \in [-20, 20]$. To ensure the feasibility of the problem, we take $b = Al + rA(u - l)$, where $l = (1, 1, \dots, 1)^T$, $u = (5, 5, \dots, 5)^T$ and $r = 0.7$.

In the implementation, the quadratic continuous problem (P') is solved by using Lemke's complementary pivoting algorithm, see [1]. Our numerical results are summarized in Tables 1, 2, 3, and 4, where *min*, *max*, and *avg* stand for the minimum, maximum and average results by running the algorithm for ten test problems for each size n and m.

Table 1 Numerical results for constraints $m = 10$

	CPU time (s)			Number of sub-boxes			Number of iterations		
n	Min	Max	Avg	Min	Max	Avg	Min	Max	Avg
10	0.016	0.297	0.123	54	593	284.6	8	92	42.0
15	0.078	1.734	0.605	61	3301	846.3	5	395	92.1
20	0.172	4.828	1.278	207	4746	1143.3	17	473	105.8
30	0.016	59.188	12.605	1	32,716	6417.8	1	2536	462.2
40	1.734	2018.797	400.572	79	608,094	117810.5	1	29,178	5638.4
50	11.734	5463.391	1576.763	925	1,104,924	277602.3	31	42,590	10629.5
60	14.313	5171.297	965.266	378	825,914	122427.6	6	32,550	4568.5
65	15.031	9055.141	1087.900	586	907,556	104632.4	14	27,651	3186.8

Table 2 Numerical results for constraints $m = 15$

	CPU time (s)			Number of sub-boxes			Number of iterations		
n	Min	Max	Avg	Min	Max	Avg	Min	Max	Avg
10	0.094	0.844	0.313	95	1424	559.8	14	259	93.8
15	0.797	37.750	11.666	608	44,865	14454.1	66	6423	1918.1
20	1.313	18.922	5.822	609	13,194	3836.1	40	1167	343.1
30	3.375	309.766	105.188	860	154,318	43438.9	60	11,020	2873.5
40	6.422	2209.750	413.711	920	397,227	79967.1	30	16,532	3587.8
45	33.969	4133.125	1243.214	5364	715,387	227297.9	237	30,331	10273.4

Table 3 Numerical results for constraints $m = 20$

	CPU time (s)			Number of sub-boxes			Number of iterations		
n	Min	Max	Avg	Min	Max	Avg	Min	Max	Avg
10	0.094	2.000	0.991	77	2966	1015.1	11	596	183.9
15	0.281	168.609	26.052	97	109,113	17938.6	7	13,869	2226.2
20	3.344	88.375	38.875	1803	53,386	20348.3	139	6006	1956.2
25	2.250	4052.078	526.523	590	1,751,016	219537.8	35	155,382	18839.5
40	117.563	16961.469	3763.525	22,873	2,819,007	601553.2	1120	141,560	29133.0

Table 4 Numerical results for constraints $m = 25$

	CPU time (s)			Number of Sub-boxes			Number of Iterations		
n	Min	Max	Avg	Min	Max	Avg	Min	Max	Avg
10	0.984	18.391	5.308	85	1658	649.4	13	324	115.8
15	2.109	108.297	29.972	537	54,237	16028.8	56	6860	2020.4
20	1.578	1058.609	199.477	556	488,854	84138.5	58	50,855	8455.3
25	26.766	13231.578	3035.883	8396	3,929,736	903208.0	634	368,027	78571.5

From the results in Tables 1, 2, 3, and 4, we can observe that the algorithm can find the exact optimal solution of nonseparable concave quadratic integer programming problems in reasonable computation time. The algorithm is efficient considering the difficulty caused by the nonseparability, concavity, and the discrete nature of the problem. The efficiency of the algorithm relies on reducing the optimality gap gradually by the linear and contour cut technique. According to our computational experiments, the CPU time used by the algorithm depends both on the number of sub-boxes and the computation time to search for the optimal solution to the convex quadratic programming problem in each sub-box. Therefore, the algorithm will be more efficient if the lower bound can be found fast and further improved.

The performance of Algorithm 4.1 has been compared with the traditional branch and bound method based on the quadratic convex relaxation problem (P'). The comparison results are reported in Table 5, where average CPU time, average sub-box number (or average branches), and average iterations are obtained by running ten test problems for each size.

Table 5 Comparison results with the traditional branch and bound method

	Algorithm 4.1			Traditional BB		
$n \times m$	Avg CPU(s)	Avg iters	Avg boxes	Avg CPU(s)	Avg iters	Avg branches
10×5	0.017	11.3	76.1	49.023	937.0	936.0
12×5	0.047	50.8	295.9	291.247	4196.6	4195.6
10×7	0.042	37.3	228.8	140.163	2649.8	2648.8
12×7	0.066	50.1	355.5	915.050	13132.2	13131.2

From Table 5, we can see that Algorithm 4.1 is more efficient than the traditional branch and bound method in term of average CPU time, average number of iterations, and average number of sub-boxes (branches). Algorithm 4.1 outperforms the traditional branch and bound method mainly due to the special domain cut technique adopted in the proposed algorithm. The domain cut technique removes regions that do not contain the optimal solution to the original problem, thus the feasible region can be reduced greatly, consequently we can find the optimal solution to the problem more quickly. The comparisons also show that the presented algorithm is efficient and reliable.

6 Conclusion

A new and simple exact algorithm is proposed for nonseparable concave quadratic integer programming problems in this paper. The method combines a domain cut and partition strategy with a convex relaxation problem. We obtain lower bounds of the subproblems by solving the convex relaxation problems and the optimality gap between the upper bound and the lower bound is reduced greatly by cutting off the regions not containing better feasible solution than the incumbent solution. Thus, we find the optimal solution of the primal problem in a finite number of steps. The efficiency of the proposed algorithm can be observed from our computation experiments. Also, the comparison results with the traditional branch and bound method in Table 5 further show the efficiency of the presented algorithm.

References

1. Bazaraa, M.S., Sherali, H.D., Shetty, C.M.: Nonlinear Programming: Theory and Algorithms. Wiley, New York (1993)
2. Beck, A., Teboulle, M.: Global optimality conditions for quadratic optimization problems with binary constraints. SIAM J. Optimiz. **11**, 179–188 (2000)
3. Benson, H.P., Erengue, S.S.: An algorithm for concave integer minimization over a polyhedron. Nav. Res. Log. **37**, 515–525 (1990)

4. Bretthauer, K.M., Shetty, B.: The nonlinear resource allocation problem. Oper. Res. **43**, 670–683 (1995)
5. Bretthauer, K.M., Shetty, B.: The nonlinear knapsack problem-algorithms and applications. Eur. J. Oper. Res. **138**, 459–472 (2002a)
6. Bretthauer, K.M., Shetty, B.: A pegging algorithm for the nonlinear resource allocation problem. Comput. Oper. Res. **29**, 505–527 (2002b)
7. Cabot, A.V., Erengue, S.S.: A branch and bound algorithm for solving a class of nonlinear integer programming problems. Nav. Res. Log. **33**, 559–567 (1986)
8. Guignard, M., Kim, S.: Lagrangian decomposition: a model yielding stronger lagrangian relaxation bounds. Math. Program. **33**, 262–273 (1987)
9. Hochbaum, D.: A nonlinear knapsack problem. Oper. Res. Lett. **17**, 103–110 (1995)
10. Horst, R., Tuy, H.: Global Optimization: Deterministic Approaches. Springer, Heidelberg (1993)
11. Ibaraki, T., Katoh, N.: Resource Allocation Problems: Algorithmic Approaches. MIT Press, Cambridge, Mass (1988)
12. Kodialam, M.S., Luss, H.: Algorithm for separable nonlinear resource allocation problems. Oper. Res. **46**, 272–284 (1998)
13. Lasserre, J.B.: An explicit equivalent positive semidefinite program for nonlinear 0–1 programs. SIAM J. Optimiz. **12**, 756–769 (2002)
14. Li, D., Sun, X.L., Wang, F.L.: Convergent Lagrangian and contour cut method for nonlinear integer programming with a quadratic objective function. SIAM J. Optimiz. **17**, 372–400 (2006)
15. Li, D., Sun, X.L., Wang, J., McKinnon, K.: Convergent Lagrangian and domain cut method for nonlinear knapsack problems. Comput. Optim. Appl. **42**, 67–104 (2009)
16. Marsten, R.E., Morin, T.L.: A hybrid approach to discrete mathematical programming. Math. Program. **14**, 21–40 (1978)
17. Mathur, K., Salkin, H.M., Morito, S.: A branch and search algorithm for a class of nonlinear knapsack problems. Oper. Res. Lett. **2**, 55–60 (1983)
18. Michelon, P., Maculan, N.: Lagrangian decomposition for integer nonlinear programming with linear constraints. Math. Program. **52**, 303–313 (1991)
19. Michelon, P., Maculan, N.: Lagrangian methods for 0–1 quadratic programming. Discre. Appl. Math. **42**, 257–269 (1993)
20. Pardalos, P.M., Rosen, J.B.: Reduction of nonlinear integer separable programming problems. Int. J. Comput. Math. **24**, 55–64 (1988)
21. Sun, X.L., Li, D.: Optimality condition and branch and bound algorithm for constrained redundancy optimization in series systems. Optim. Eng. **3**, 53–65 (2002)
22. Sun, X.L., Wang, F.L., Li, D.: Exact algorithm for concave knapsack problems: Linear underestination and partition method. J. Global Optim. **33**, 15–30 (2005)
23. Wang, F.L., Sun, X.L.: A Lagrangian decomposition and domain cut algorithm for nonseparable convex knapsack problems. Oper. Res. Trans. **8**, 45–53 (2004)

Zeitfracht Medien GmbH
Ferdinand-Jühlke-Straße 7
99095 Erfurt, Deutschland
produktsicherheit@kolibri360.de